ENVIRONMENTAL SCIENCE, ENGINEERING AND TECHNOLOGY

CASE APPLICATIONS OF STATISTICAL MODELS IN ECOLOGY AND EVOLUTION

ENVIRONMENTAL SCIENCE, ENGINEERING AND TECHNOLOGY

Additional books in this series can be found on Nova's website under the Series tab.

Additional e-books in this series can be found on Nova's website under the e-book tab.

ENVIRONMENTAL SCIENCE, ENGINEERING AND TECHNOLOGY

CASE APPLICATIONS OF STATISTICAL MODELS IN ECOLOGY AND EVOLUTION

YOUHUA CHEN

nova publishers
New York

Library of Congress Cataloging-in-Publication Data

Names: Chen, Youhua (Biologist), editor.
Title: Case applications of statistical models in ecology and evolution / Youhua Chen (University of Alberta, Edmonton, Canada), editor.
Description: Hauppauge, New York : Nova Science Publishers, Inc., [2016] | Series: Science, evolution, and creationism | Includes bibliographical references and index. | Description based on print version record and CIP data provided by publisher; resource not viewed.
Identifiers: LCCN 2016017775 (print) | LCCN 2016007432 (ebook) | ISBN 9781634848947 () | ISBN 9781634848763 (hardcover)
Subjects: LCSH: Ecology--Mathematical models. | Evolution (Biology)--Mathematical models. | Biogeography--Mathematical models.
Classification: LCC QH541.15.M3 (print) | LCC QH541.15.M3 C364 2016 (ebook) | DDC 577.01/51--dc23
LC record available at https://lccn.loc.gov/2016017775

Published by Nova Science Publishers, Inc. † New York

CONTENTS

Preface		**vii**
Part 1: Case Applications		**1**
Chapter 1	Permutation-Free Fisher's Z-Transform Method for Significance Testing of Mantel's Statistic	**3**
Chapter 2	Random Species/Population Change and Multiple-Site Beta Diversity Indices: A Comparison of Contradictory Scenarios from Simulation- and Empirical BCI Tree-Derived Data Sets	**11**
Chapter 3	On the Relationship between Community-Level Beta Diversity and Partitioning Variation for Space and Environment: BCI Tree Community as a Case Study	**29**
Chapter 4	Distinguishing Niche and Neutral Processes: Issues in Variation Partitioning Statistical Methods and Further Perspectives	**41**
Chapter 5	Estimating Ancestral Ranges in Historical Biogeography: On the Development of Alternative Range Evolution Models for Maximum Likelihood Inference	**55**
Chapter 6	On the Relationships of Clade Age, Species Richness, and Phylogenetic Diversity of Global Mammals	**67**

Chapter 7 Spatial Point Pattern Analysis of County Locations of Mainland China: The Relative Roles of Anthropogenic Factors, Climatic Fluctuation and Water Availability **75**

Chapter 8 Modeling Species-Area Relationship with Measurement Uncertainty **87**

Chapter 9 Estimating Population Change of Species under Climate Change Using Occupancy-Abundance Relationship **97**

Chapter 10 'Coexist': An R Package for Performing Species Coexistence Modeling and Analysis **111**

Chapter 11 Fitness Equivalence Promotes Species Coexistence under Fluctuating Environments with Simultaneous Coupling of Dispersal and Competition **119**

Part 2: Appendices of R Codes Used for Computation and Simulation **139**

Appendix 1 R Code for Fisher Transform and Simulation **141**

Appendix 2 R Code for Simulating Asymmetric Dispersal and Coexistence **145**

Author's Contact Informaion **193**

Index **195**

PREFACE

Biodiversity conservation is of primary importance to build up a harmonious relation between human beings and the Nature. Biodiversity conservation requires comprehensive inventories of biodiversity resources, in-depth analyses of biodiversity data, and wise understanding of biodiversity build-up mechanisms through the interplay of ecology and evolutionary biology. Nonetheless, elucidating the pertinent ecological and evolutionary processes in biodiversity maintenance is not an easy task as it requires critical thinking and statistical tools to help extract informative results and patterns from complex ecological and biological data. As such, application of appropriate statistical methods is very important for ecologists and evolutionary biologists.

In this book, I present some case applications of statistical methods for the purpose to provide readers of the book some senses on how to apply statistical models and methods in ecological and evolutionary studies. The cases I arranged in this book seem a bit messy, but I made such an arrangement in purpose: I hope the readers to jump into the jungle of ecological and evolutionary researches with complex statistical methods, I do not expect the readers to fully understand these methods in a single book I wrote. I show them to the readers with very simple or even silly examples and data sets. As long as my book can help readers to roughly know what kinds of statistical methods or knowledge should be equipped when addressing specific research questions in ecology and evolution, my hard work on writing it is half the battle.

This work is funded by the China Scholarship Council, who allows me to have a chance to study abroad. I always appreciate my parents for their endless loves and supports, in particular during the hard time I faced.

Youhua Chen
2016-01-26

PART 1: CASE APPLICATIONS

Chapter 1

PERMUTATION-FREE FISHER'S Z-TRANSFORM METHOD FOR SIGNIFICANCE TESTING OF MANTEL'S STATISTIC

ABSTRACT

Fisher's Z-transform method has been adopted to test the significance of Mantel statistic (or broadly, Pearson's correlation statistic) from randomness. The statistical method is totally permutation-free and thus the calculation is much faster. The method thus is much superior for situation when one compared two large or many distance matrices simultaneously. In conclusion, the present study suggests an easy but fast statistical method to quantify the significance level of observed Mantel statistic when comparing two distance matrices.

Keywords: Mantel statistic, matrix comparison, Z-transform, asymptotic normal distribution, community ecology

INTRODUCTION

Mantel statistic (Mantel, 1967; Sokal, 1986) has been widely used to infer the associations between geographic distances, genetic distances or compositional distances in population and community ecology (Legendre and Fortin, 2010; Wollenberg et al., 2011; Borcard and Legendre, 2012; Crottini et al., 2012). The significance of observed Mantel statistic derived from two distance matrices was tested typically by carrying out the permutations of the data (Legendre and Legendre, 1998). Such a standard protocol has been integrated into many available programs or software. For example, *vegan* (Oksanen et al., 2012), *Arlequin* (Excoffier and Schneider, 2005) and so on.

However, the permutation process requires a lot of computational loads, which is not feasible especially when one compared two large distance matrices. As such, utilizing statistical methods to accurately infer the significant difference of observed Mantel statistic in comparison to randomly simulated ones should be advantageous since it is totally simulation-free and the calculation is much faster.

In the present study, we adopted Fisher's Z-transformation (namely as Fisher's method hereafter) (Fisher, 1921; Hardle and Simar, 2012) to test the significance for Mantel statistic (or broadly, Pearson's correlation statistic). Fisher's method has been well introduced in many multivariate statistical textbooks for testing the significance of correlations (Hardle and Simar, 2012). Surprisingly, it has not been utilized so far among ecologists.

For comparison, we also performed a series of numerical simulations to show the relationship between Mantel statistics and associated statistical significances derived from both permutation test and Fisher's method. The purpose of the study is to propose Fisher's Z-transform method to estimate probability of significance difference based on the observed Mantel statistic and normal distribution without carrying out any permutation tests.

MATERIALS AND METHODS

Mantel Test

The standardized Mantel statistic was calculated as follows (Mantel, 1967; Legendre and Legendre, 1998). For two distance matrices X and Y,

$$r_M = \frac{1}{D-1}\sum_{i=1}^{n-1}\sum_{j=i+1}^{n}\left(\frac{x_{ij}-\bar{x}}{s_x}\right)\left(\frac{y_{ij}-\bar{y}}{s_y}\right) \tag{1}$$

Where $D = n(n-1)/2$ is the number of elements in the lower triangular part of each matrix. $\bar{x}$ and $\bar{y}$ are the means of the values in each of the lower triangular parts from *X* and *Y*. s_x and s_y are their standard deviations.

Fisher's Z-transformation method

$$W = \frac{1}{2}\log(\frac{1+r_M}{1-r_M}) \tag{2}$$

Under the null hypothesis, r_M is not significantly different from $r_0 = 0$, as such, W followed an asymptotic normal distribution (Fisher, 1921; Hardle and Simar, 2012). Then, the expectation and variance of W statistic were given by (Hardle and Simar, 2012),

$$E(W) = \frac{1}{2}\log(\frac{1+r_0}{1-r_0}) = 0 \tag{3}$$

$$Var(W) = \frac{1}{D-3} \tag{4}$$

It could be found that,

$$Z = \frac{W - E(W)}{\sqrt{Var(W)}} = \frac{W}{\sqrt{Var(W)}} = \sqrt{(D-3)} \times W \sim N(0,1) \tag{5}$$

Which is directly from Theorem 3.2 in page 80 of the book (Hardle and Simar, 2012).

For a unit normal distribution, the 2.5% and 97.5% quantiles were -1.96 and 1.96 respectively. As a consequence, when $Z \geq 1.96$ or $Z \leq -1.96$, we could ascertain that the observed Mantel statistic is significantly different from the expected random ones with $P \leq 0.05$. For the higher significant level $P \leq 0.001$, the corresponding Z values are $Z \geq 3.29$ (99.95% quantile) or $Z \leq -3.29$ (0.05% quantile).

For small sample sizes (e.g., $D \leq 25$), we could have a Hotelling's transform (Hotelling, 1953; Hardle and Simar, 2012) for the correction of sampling biases as,

$$W^* = W - \frac{3W + \tanh(W)}{4(D-1)} \tag{6}$$

here,

$$Var(W^*) = \frac{1}{D-1} \tag{7}$$

So, the corrected Z value for small data sets reads,

$$Z^* = \frac{W^* - E(W^*)}{\sqrt{Var(W^*)}} = \sqrt{(D-1)} \times W^* \sim N(0,1) \tag{8}$$

For Morran's I index for measuring spatial autocorrelation, the *Z*-formula for measuring significance is given by,

Permutation Test

For carrying out permutation test, I permuted all the rows of one matrix while keeping the other unchanged. As such, I could obtain a hypothetical Mantel statistic from a simulation.

Numerical Simulations

I designed six treatments, in which the row number of one matrix was fixed to be [30, 50, 100, 200, 300, 500]. The corresponding column number of that matrix was set to randomly choose the ranges [20~40, 40~60, 50~150, 150~250, 250~350, 450~550]. The same matrix generation strategy is applied to another matrix. As such, we could construct two matrices, which have the same row sizes, but varying column sizes. Each treatment has 500 replicates. The simulation results were presented in Figure 1.

The calculation of Mantel statistic and its associated significance was done using the function "*mantel()*" in *vegan* package (Oksanen et al., 2012) under R environment (R Development Core Team, 2013). The calculation of Fisher's Z-transform statistic was done by an R script written by the authors.

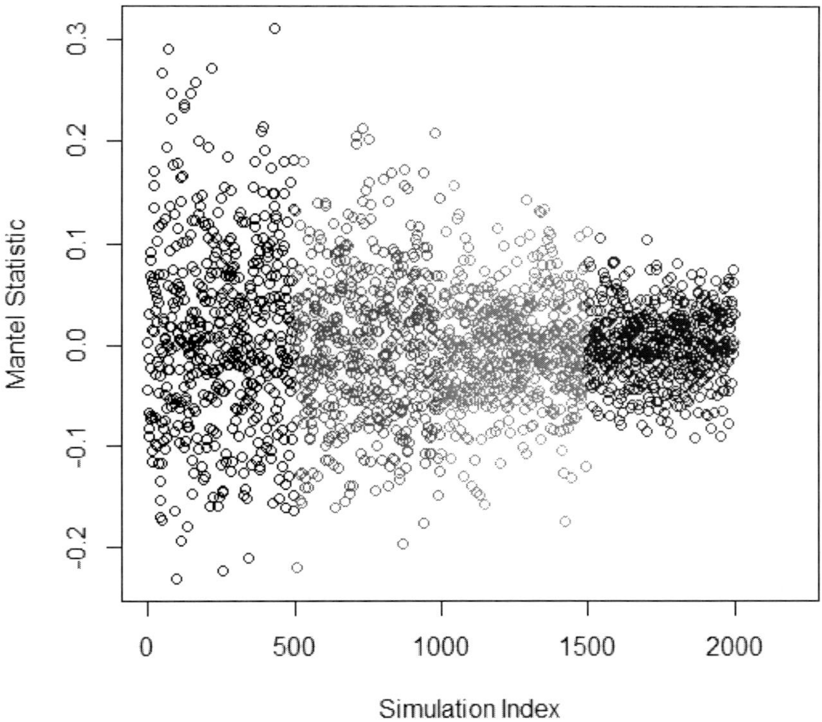

Figure 1. The relationship between matrix size and Mantel statistic. Each 500 indices in the X-axis indicated a treatment.

RESULTS AND DISCUSSION

As showed by the simulation results (Figure 2), there is a closely negative relationship between the *P* values and *Z* scores of the Mantel statistics. Higher *P* values derived from permutation tests will always result into low *Z* scores, indicating that the probability of accepting null hypothesis (the observed Mantel statistic is not different from 0) is increasing. In contrast, lower P values would be accompanied with high *Z* scores.

As such, the utilization of Fisher's Z-transform statistic could allow ones to test the significance of standardized Mantel statistic rigorously without performing computation-intensive permutations. Fisher's Z-transform thus is recommended to test and compare the significance of Mantel statistics for the situations when large or many datasets are required to test simultaneously.

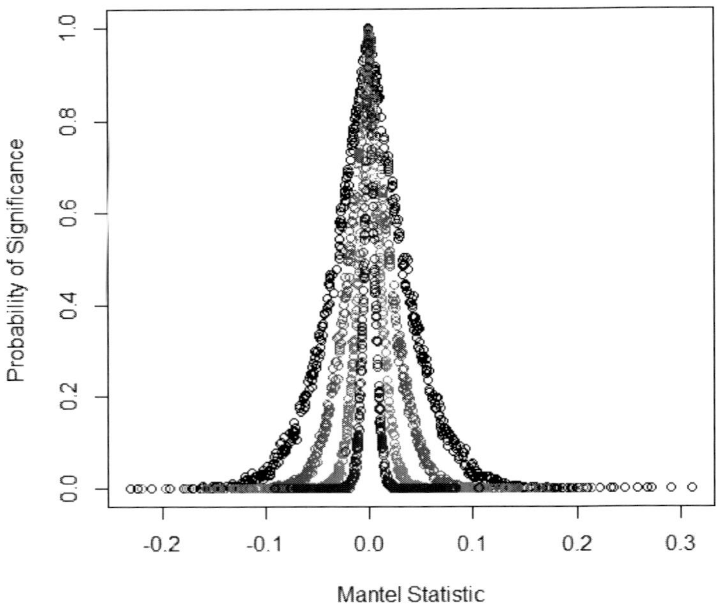

Figure 2. The relationship between Mantel statistic and probability of accepting null hypothesis.

In conclusion, the present note showed that Z-transformation could offer a very simple way to test the significance of Mantel test, which is broadly used in ecological studies for comparing species-site matrices and testing hypotheses relevant to beta diversity patterns in community ecology.

ACKNOWLEDGMENTS

This work is supported by the China Scholarship Council.

REFERENCES

Borcard, D., and P. Legendre (2012). Is the Mantel correlogram powerful enough to be useful in ecological analysis? A simulation study. *Ecology* 93, 1473–1481.

Crottini, A., J. Brown, V. Mercurio, F. Glaw, M. Vences, and F. Androne (2012). Phylogeography of the poison frog Mantella viridis (Amphibia:

Mantellidae) reveals chromatic and genetic differentiation across ecotones in northern Madagascar. *J. Zool. Syst. Evol. Res.* 50, 305–314.

Excoffier, L., and S. Schneider (2005). Arlequin ver. 3.0: An integrated software package for population genetics data analysis. *Evol. Bioinform.* Online 1, 47–50.

Fisher, R. (1921). On the "probable error" of a coefficient of correlation deduced from a small sample. *Metron* 1, 1–32.

Hardle, W., and L. Simar (2012). Applied multivariate statistical analysis, *Springer* Berlin Heidelberg.

Hotelling, H. (1953). New light on the correlation coefficient and its transformation. *J. R. Stat. Soc. Ser. B* 15, 193–232.

Legendre, P., and M. Fortin (2010). Comparison of the Mantel test and alternative approaches for detecting complex multivarite relationships in the spatial analysis of genetic data. *Mol. Ecol. Resour.* 10, 831–844.

Legendre, P., and L. Legendre (1998). Numerical ecology, *Elsevier Science* BV.

Mantel, N. (1967). The detection of disease clustering and a generalized regression approach. *Cancer Res.* 27, 209–220.

Oksanen, J., G. Blanchet, R. Kindt, and et al. (2012). vegan: Community Ecology Package. *R package version* 2.0-4.

R Development Core Team (2013). R: A Language and Environment for Statistical Computing, Vienna, Austria. ISBN 3-900051-07-0, URL http://www.R-project.org.

Sokal, R. (1986). Spatial data anlaysis and historical processes, p. 29–43. In E. Diday, Y. Escoufier, L. Lebart, J. Pages, Y. Schektman, and R. Tomassone [eds.], Data analysis and informatics, IV. North-Holland, Amsterdam, The Netherlands.

Wollenberg, K., D. Vieites, F. Glaw, and M. Vences (2011). Speciation in little: the role of range and body size in the diversification of Malagasy mantellid frogs. *BCM Evol. Biol.* 11, 217.

Chapter 2

RANDOM SPECIES/POPULATION CHANGE AND MULTIPLE-SITE BETA DIVERSITY INDICES: A COMPARISON OF CONTRADICTORY SCENARIOS FROM SIMULATION- AND EMPIRICAL BCI TREE-DERIVED DATA SETS

ABSTRACT

Which multiple-site diversity indices have the power to reveal the changes of species or population caused by habitat fragmentation and loss? Are there any differences of these multiple-site indices on reflecting the species or population changes? Several multiple-site biodiversity indices were calculated for comparison, including community turnover, nestedness, richness difference, replacement, and WNODF index. I performed both simulation and empirical studies on testing the behaviors of these indices in responses to biodiversity change. For numerical simulations, I constructed random species-site abundance matrices. For an empirical case study, I utilized BCI tree data set and also partitioned the whole data set into small data sets for a comprehensive investigation. To mimic biodiversity change process due to habitat alternation and loss, I considered three random simulation strategies: The first one is to move the individuals of some species from sites with high abundance to those without the presence of the species but keeping the total abundance of each species unchanged (named as homogenizing process). The second biodiversity change strategy is to allow some of individuals from a part of species to be reduced while the same number of individuals of another part of species to be added. This process could keep the total abundance of all species in the community unchanged (named as removing/adding process). The third strategy is to delete some of the populations of certain species in the community. I randomly removed some individuals of given species to make them become rare in the community, so the total abundance of species is apparently decreasing when the process is happening (named as deleting process). This strategy can also allow the deletion of species from the community when their populations are completely removed. These three strategies were applied to both simulated-based and BCI tree-based data sets. Random species/population change in pseudo-communities which were generated from random perfectly nested matrices could remarkably change the turnover and nestedness indices, but the other indices did not vary a lot. However, an empirical study using BCI tree datasets showed contradictory consequences. Species/population change could lead to remarkable change of replacement and richness difference indices for BCI community data sets, while nestedness and turnover indices only changed slightly. Furthermore, some interesting patterns emerged from BCI data sets: I found that species/population deleting process could decrease community nestedness and community richness difference indices, while increase community replacement index. These patterns were totally unexpected. A possible reason for this unexpected pattern might be due to the deleting process could allow the removal of species which populations were totally removed and sites where no species were found anymore. Overall, I found that WNODF index did not change considerably in both cases of pseudo-communities and BCI data sets. All the multiple-site beta-diversity indices (except WNODF index) could be used comparatively so as to better quantify compositional change or species/population loss/gain in ecological communities.

Keywords: biodiversity loss, species turnover and nestedness, multiple-site comparison, simulation, BCI tree diversity

INTRODUCTION

Impacts of biodiversity loss on community structure have been widely studied. For instance, species extinction could disrupt ecosystem functioning (Crowe et al. 2011; Larsen et al. 2005), affect ecological patterns of diseases and vectors (Pongsiri et al. 2009) and so on. However, there are still a lot of gaps not being well explored. For example, if species/population changes could lead to the modification of community structure, could I quantify the effects of species/population changes using beta diversity indices accordingly? In the present study, I quantitatively assessed whether random species/population changes in communities could be reflected by different multiple-site biodiversity indices.

Why do I need to assess the impacts of species/population change on modifying beta-diversity patterns of the community? It can direct us to uncover the relationships between community beta diversity structure and community stability, measure the effectiveness of beta diversity indices on quantifying the community compositional variation. Currently, it is still quite unclear about the effectiveness of multiple-site biodiversity indices on capturing species/population fluctuation in the community. Limited evidences have been found in previous works (Almeida-Neto et al. 2011, 2012; Baselga 2012; Carvalho 2012a, b), but not systematic.

Random species/population change should be a natural phenomenon in ecological communities. However, there is no link between random species/population change and beta diversity indices. As stated above, if I could be able to quantify random species/population change in the community using available multiple-site beta diversity indices, I could be able to better reveal community variation by simple indices' values. Some previous works have stated that a good beta diversity index should have explicit biological meanings, and can measure the change of beta diversity in the community sensitively (Carvalho 2012a, b). Using simulation-driven and BCI tree empirical data sets, I could quantify the effectiveness of different kinds of multiple-site beta diversity indices on capturing community beta diversity while holding relevant biological meanings.

In a summary, in the present study, I provided a relatively comprehensive comparison of different multiple-site beta diversity indices and discussed the potential powers of each index as the indicator of species/population changes in the community using theoretical and empirical data sets.

MATERIALS AND METHODS

Partitioning of Community Beta-Diversity

Beta-diversity for multiple sites (i.e., community) could be partitioned into two components, turnover and nestedness (Baselga et al. 2007; Baselga 2010). In recent works, beta-diversity might be better to partition into replacement and richness difference so as to be ecologically meaningful (Carvalho et al. 2012a,b). I have developed multiple-site replacement and richness difference indices. Thus, I chose several multiple-site indices as bellows to quantify community-level diversity, which were still in controversy on their effectiveness and biological relevance (Almeida-Neto et al. 2012; Baselga 2012; Carvalho et al. 2012b). All the indices were calculated using an R package “MBI” developed by the author (available at the URL: http://cran.r-project).

Community Turnover

I used a beta-diversity similarity index accounting for multiple sites (Baselga et al. 2007; Baselga 2012) to measure the community turnover. The index was formulated as follows,

$$\beta_{sim} = \frac{\sum_{i<j} \min(b_{ij}, b_{ji})}{(\sum_i S_i - S_T) + (\sum_{i<j} \min(b_{ij}, b_{ji}))} \tag{1}$$

Where, S_i is the number of species found in site i, while S_T total number of species in the community. b_{ij} and b_{ji} are the number of species exclusive to sites i and j respectively when compared by pairs.

Community Nestedness

I used the nestedness index developed for multiple sites to quantify community nestedness pattern (Baselga 2012; Baselga 2010), which is written as,

$$\beta_{nes} = \frac{\sum_{i<j} \max(b_{ij}, b_{ji}) - \sum_{i<j} \min(b_{ij}, b_{ji})}{2(\sum_i S_i - S_T) + \sum_{i<j} \max(b_{ij}, b_{ji}) + \sum_{i<j} \min(b_{ij}, b_{ji})} \times \frac{(\sum_i S_i - S_T)}{(\sum_i S_i - S_T) + \sum_{i<j} \min(b_{ij}, b_{ji})} \tag{2}$$

The definitions of the letters here are identical to those in community turnover index (1).

I also considered another nestedness index developed recently (Almeida-Neto and Ulrich 2011), which has been used to compare Baselga's nestedness index (Almeida-Neto *et al.* 2012).

Community WNODF Index

The mean nestedness value for all pairs of columns is,

$$WNODFc = 100\sum_{i=1}^{n-1}\sum_{j=i+1}^{n}\frac{k_{ij}}{N_j} \tag{3}$$

Where k_{ij} denotes the number of cells with lower values in column c_j compared to those in column c_i and N_j is the total number of non-empty cells in column c_j.

Identically, the mean nestedness value for all pairs of rows is,

$$WNODFr = 100\sum_{i=1}^{m-1}\sum_{j=i+1}^{m}\frac{k'_{ij}}{N'_j} \tag{4}$$

Where k'_{ij} denotes the number of cells with lower values in row r_j compared to those in row r_i and N'_j is the total number of non-empty cells in row r_j.

The combined metric was written as,

$$WNODF = \frac{2(WNODFc + WNODFr)}{m(m-1) + n(n-1)} \tag{5}$$

Assuming that there *n* columns and *m* rows in the community matrix.

Community Replacement

I used the replacement index developed for multiple sites to quantify community replacement pattern, the deduction is quite straightforward by following the same procedure for nestedness index (Baselga 2010) as,

$$\beta_{rep} = 2 \times \frac{\sum_{i<j} \min(b_{ij}, b_{ji})}{\left(\sum_{i} S_i - S_T\right) + \sum_{i<j} (b_{ij} + b_{ji})} \tag{6}$$

Community Richness Differences

I used the richness difference index developed for multiple sites to quantify community richness difference pattern, the deduction is quite straightforward following the same procedure for turnover index:

$$\beta_{rich} = \frac{\sum_{i<j} \max(b_{ij}, b_{ji}) - \sum_{i<j} \min(b_{ij}, b_{ji})}{2(\sum_{i} S_i - S_T) + \sum_{i<j} \max(b_{ij}, b_{ji}) + \sum_{i<j} \min(b_{ij}, b_{ji})} \tag{7}$$

Simulation Study

Random Matrix Construction

Different to the methods illustrated in (Gotelli and Ulrich 2010; Ulrich and Gotelli 2007), I develop a new simple method to construct random matrix for species abundance across sites. The method firstly construct a fully perfect nestedness matrix, then the sites and species were permutated randomly to generate a reduced nested matrix, with the nestedness degrees for the whole matrix, columns and rows varying from 0 to 100 (measured by WNODF index). For simplicity, I did not take the percentage but the raw value for the whole matrix (ranging from 0 to 1) in my study.

I used this method to sample matrices with varying species and site numbers following the designed treatments as: species number=10, 30, 50, 100, 200; site number=[5,20], [20,40], [40,60], [80,120] and [100,300]. For each treatment, I sample 100 matrices and calculated the diversity indices.

Random Species/Population Changes

For assessing the roles of changing community rarity, beta-diversity and nestedness patterns which were induced by species loss, I have several strategies.

The first one is to move the individuals of some species from sites with high abundance to those without the presence of the species but keeping the total abundance of each species unchanged. This sampling procedure is running with replacement. I choose 30% of species undergoing the dispersal and relocation processes for the individuals of their own. The strategy was named as homogenizing process hereafter.

The second biodiversity change strategy is a zero-sum model, for which some of individuals from a part of species were reduced while the same number of individuals of another part of species were added. This process could keep the total abundance of all species in the community unchanged. I called it as zero-sum (removing/adding) process hereafter.

The third strategy is to delete some of the populations of certain species in the community. I randomly removed some individuals of given species to make them become rare in the community, so the total abundance of species is apparently decreasing when the process is happening. This strategy was called as deleting process hereafter. This strategy can also allow the deletion of species from the community when their populations are completely removed.

All the simulations were repeated 100 times for each of the matrices produced above.

An Empirical Test Using BCI Tree Communities

Distribution of BCI Trees in 100 Sampling Plots

The distribution of the tree abundances across the BCI sampling plots were derived from (Condit et al. 2002). There were 100 sampling sites, with 802 species in total.

For effectively utilizing this data set, I considered two means to analyze it. The first one is to utilize the whole dataset, while the second one is to partition the dataset randomly by picking up a part of total species to generate 100 sub-datasets (the sites are of course determined by the species occurrence locations) for quantitatively assessing the effects of random species on community diversity.

Results

Comparison of Different Indices Along with Population Changes in the Pseudo-Communities

Species/population changes have different consequences on the four community beta-diversity indices. As showed in Figure 1, population homogenizing process would remarkably influence community Baselga's nestedness and turnover patterns. However, homogenizing process did not remarkably change the values for the WNODF, richness difference and replacement indices, although there are slight patterns observed (e.g., WNODF and richness difference indices had a slight decreasing trend, while replacement index had a weak increasing trend).

For population zero-sum (removing/adding) and deleting processes, resultant observations were in concordance with those for homogenizing process. Baselga's nestedness and turnover indices seemed the most sensitive ones to population change in the community.

Simulation of BCI Tree Community

Interestingly, I found that the results for BCI data sets were quite different. Some new patterns emerged and were not encountered by pseudo-communities which were generated from the same rule.

As showed in Figure 4, both nestedness and richness difference indices were reduced when effect of homogenizing process is strengthened, while turnover and replacement indices were increased. However, when comparing these indices, I found that the change magnitude for richness difference (0.13~0.18) and replacement (0.8~0.85) indices is much larger than those for nestedness (0.0045~0.0065) and turnover (0.9645~0.9675) indices. These results applied to the cases for population removing/adding process (Figure 5). Thus, I found that richness difference and replacement indices were better to quantify species/population changes in high-biodiversity communities. Thus, the results from BCI data sets were incongruent with the observations from pseudo-communities.

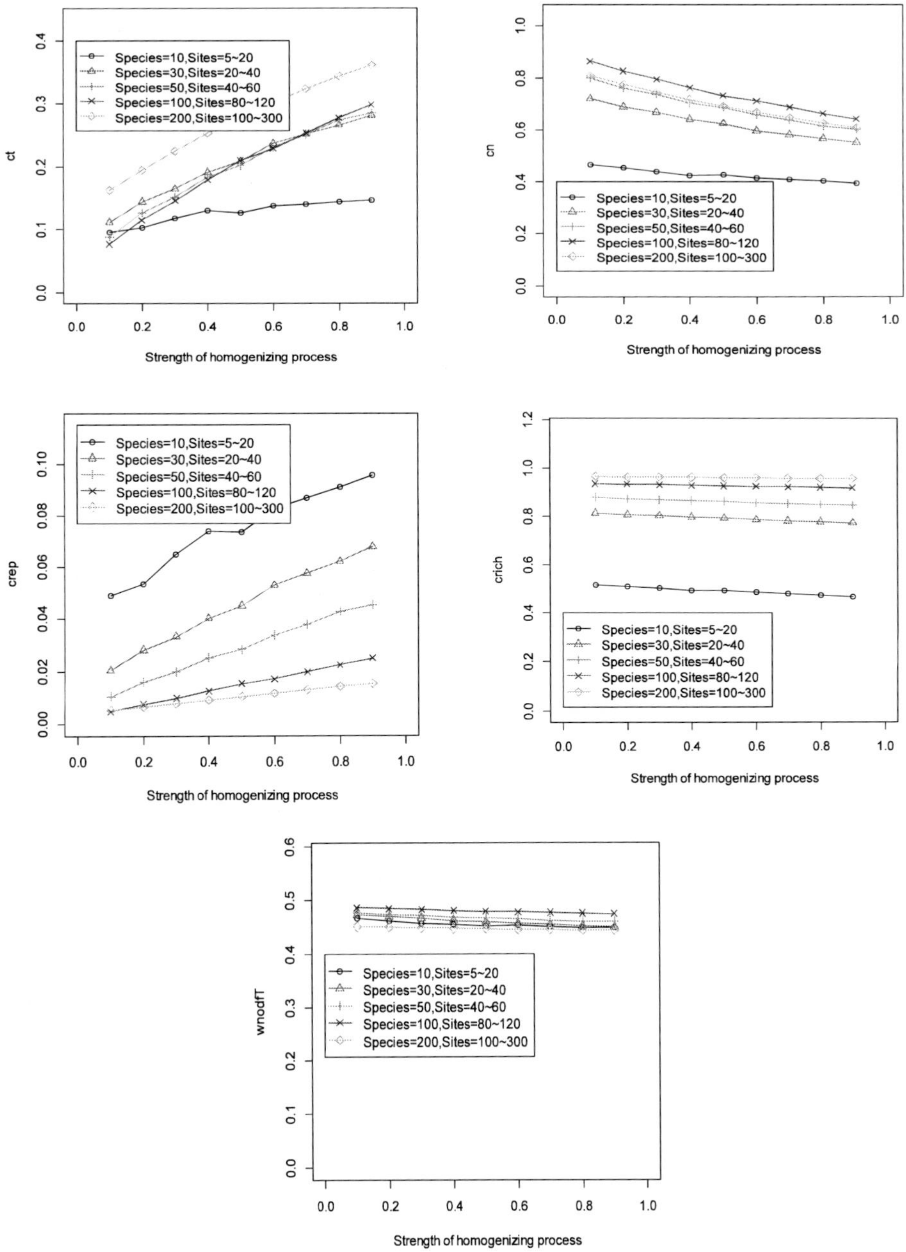

Figure 1. Effect of homogenizing process on influencing community beta diversity patterns pseudo-communities. Each line indicated the mean from 100 replicates.

WNODF index always have the decreasing trend across all the different species/population handing processes, although the decreasing magnitude is unremarkable.

Interestingly, I found that species/population deleting process could create unpredictable results. As showed in Figure 6, increasing strength of deleting effect could actually decrease community nestedness and community richness difference indices, while increase community replacement index. Community turnover index was normal as expected. This result is totally in discordance with the patterns found in population homogenizing and removing/adding processes.

One potential reason is attributed to the fact that species deleting process could allow the removal of species from the community. Species-site matrices thus have altered their row and column numbers, which should pose great impacts on community beta diversity indices.

DISCUSSION

Partitioning of beta-diversity should indicate the variation and change of community (Carvalho et al. 2012b). The turnover and nestedness indices (Baselga 2010) for pairwise comparison of sites have been found to be inconsistent with the expectations (Almeida-Neto et al. 2012; Podani and Shmera 2012; Schmera and Podani 2011).

However, in my study, I showed that different species/population change processes could be sensitively detected by nestedness and turnover indices when the original community was constructed from a fully perfect nestedness model (Figures 1-3). Other indices, WNODF, species replacement and richness difference metrics, didn't vary a lot when species handling events happened. Thus, I found that the newly proposed richness difference and replacement indices from (Carvalho et al. 2012a) and index WNODF were not as powerful as stated in previous literature on measuring random species/population change.

However, the patterns would not be held anymore for the case of empirical BCI tree community data sets. As showed in Figures 4-6, the change magnitude for community replacement and richness difference indices was the largest among the multiple-site indices. Thus, for high-biodiversity communities not derived from random perfectly nested matrices, richness difference and replacement indices seemed much powerful to capture species/population variation.

WNODF was observed not to change remarkably in both simulation- and BCI-derived data sets. This index thus should not be used as a measurement of species/population changes in the community. I believe that WNODF index should be only related to the matrix structure and property, therefore its biological meaning is not explicit.

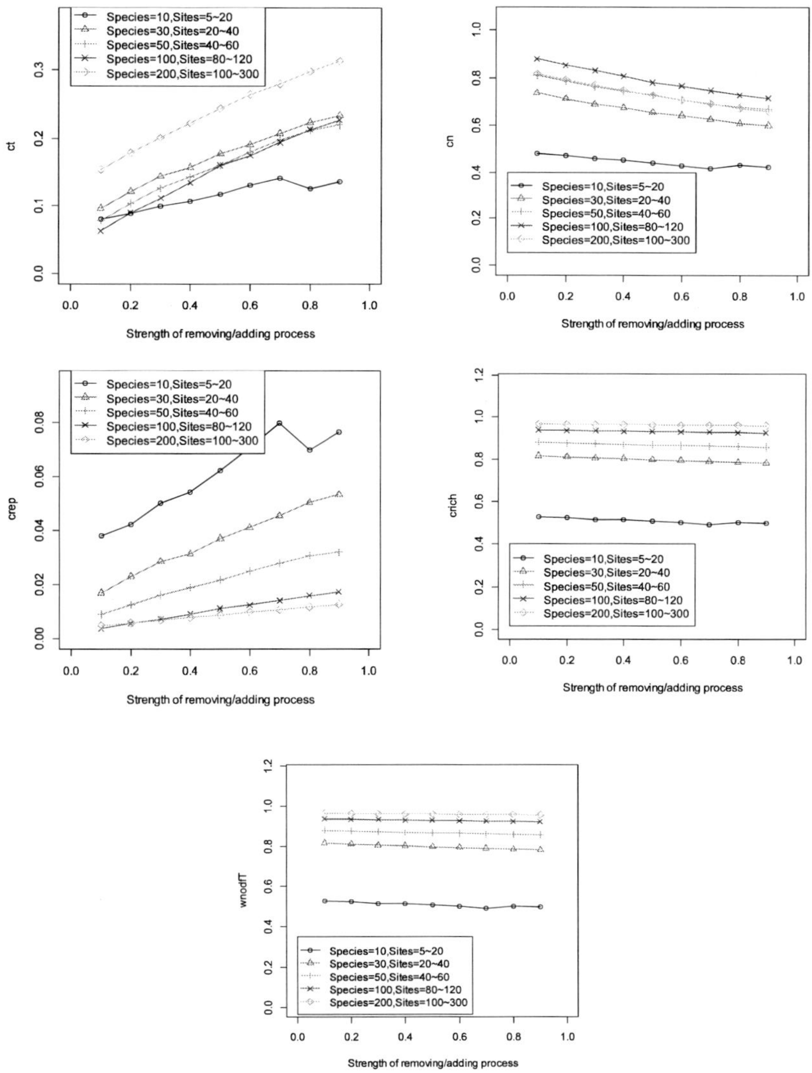

Figure 2. Effect of removing/adding process on influencing community beta diversity patterns pseudo-communities. Each line indicated the mean from 100 replicates.

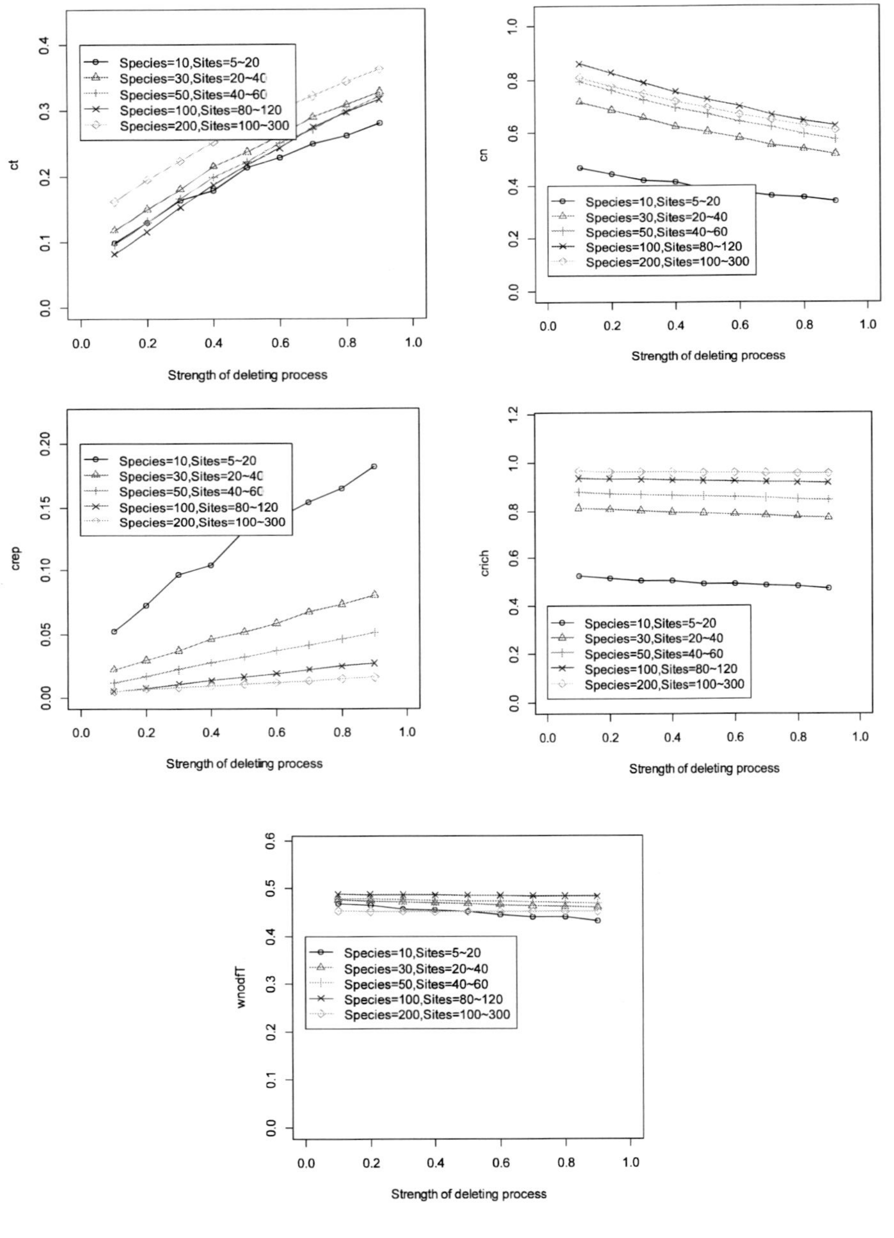

Figure 3. Effect of removing/adding process on influencing community beta diversity patterns in pseudo-communities. Each line indicated the mean from 100 replicates.

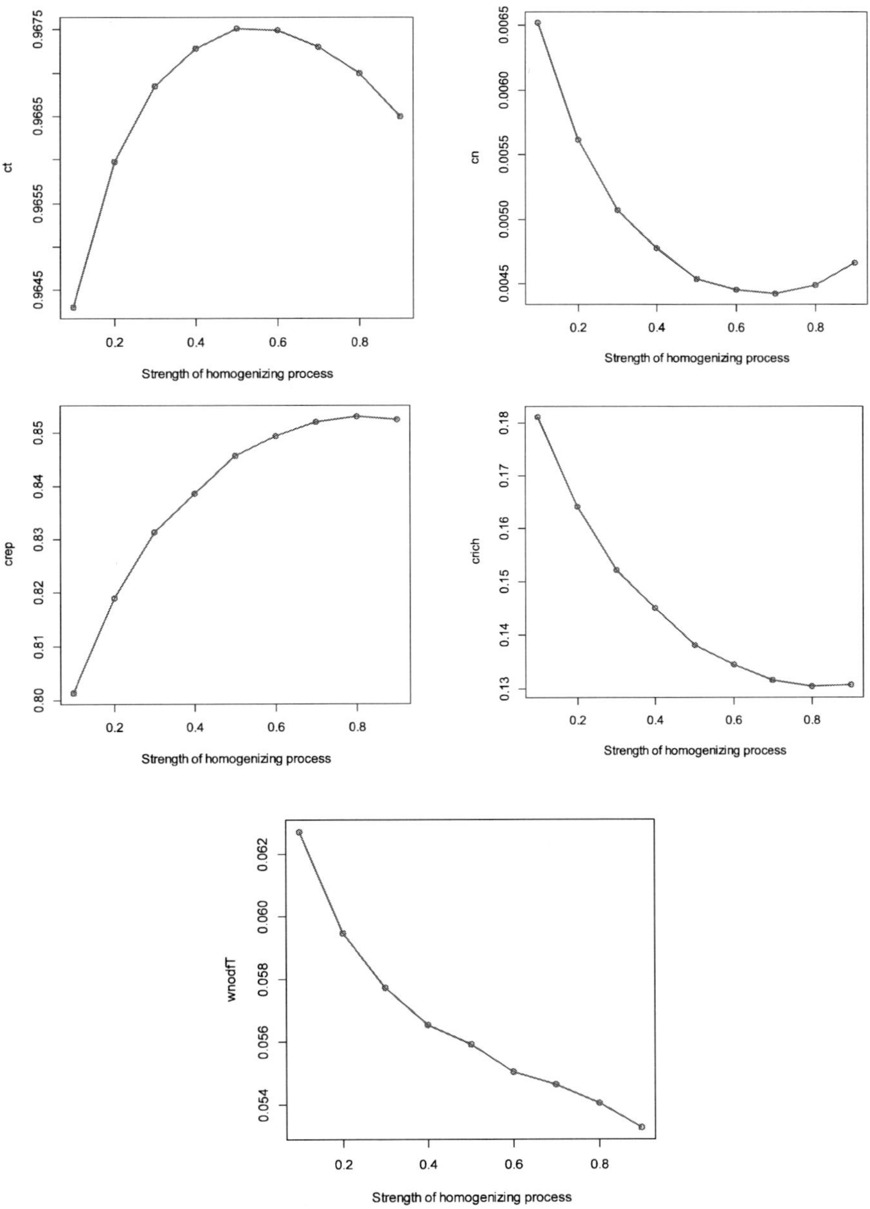

Figure 4. Effect of homogenizing process on influencing community beta diversity patterns in BCI data sets. Each line indicated the mean from 100 replicates.

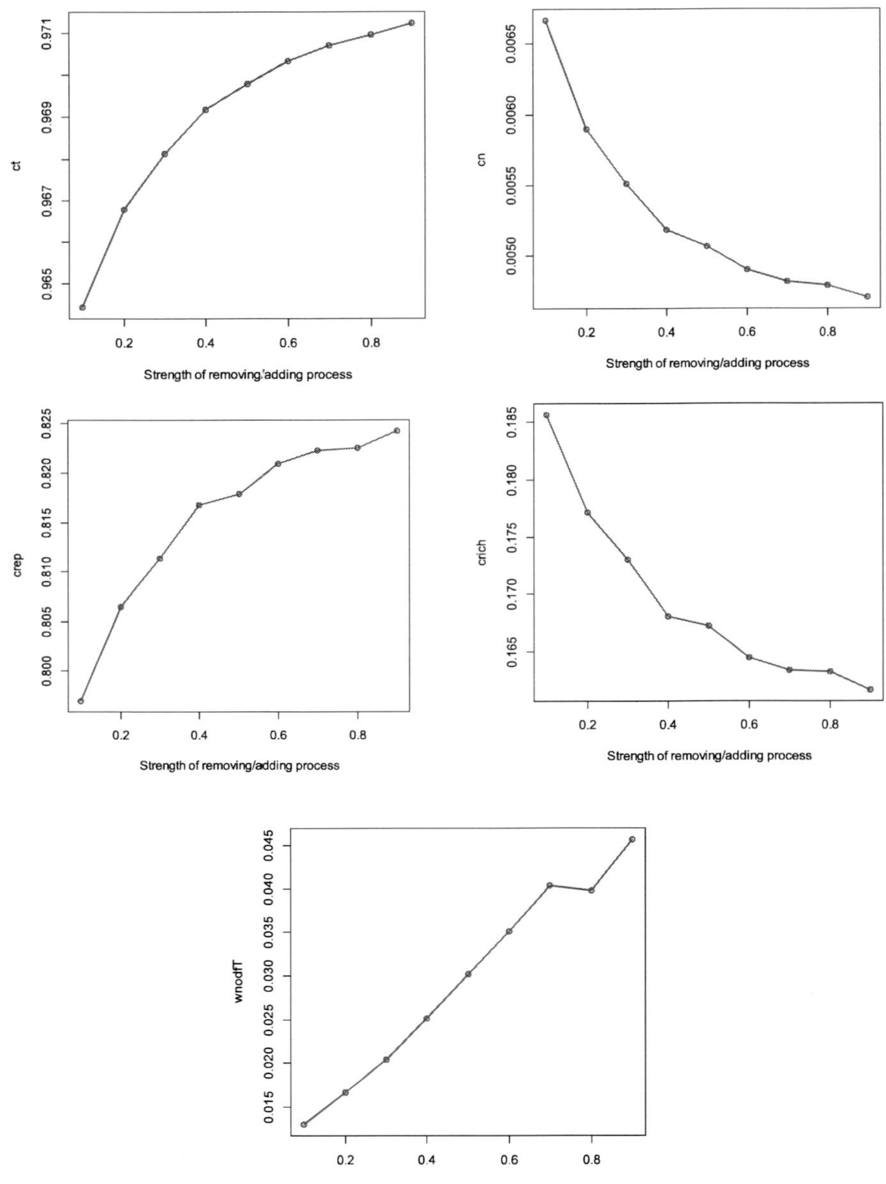

Figure 5. Effect of removing/adding process on influencing community beta diversity patterns in BCI data sets. Each line indicated the mean from 100 replicates.

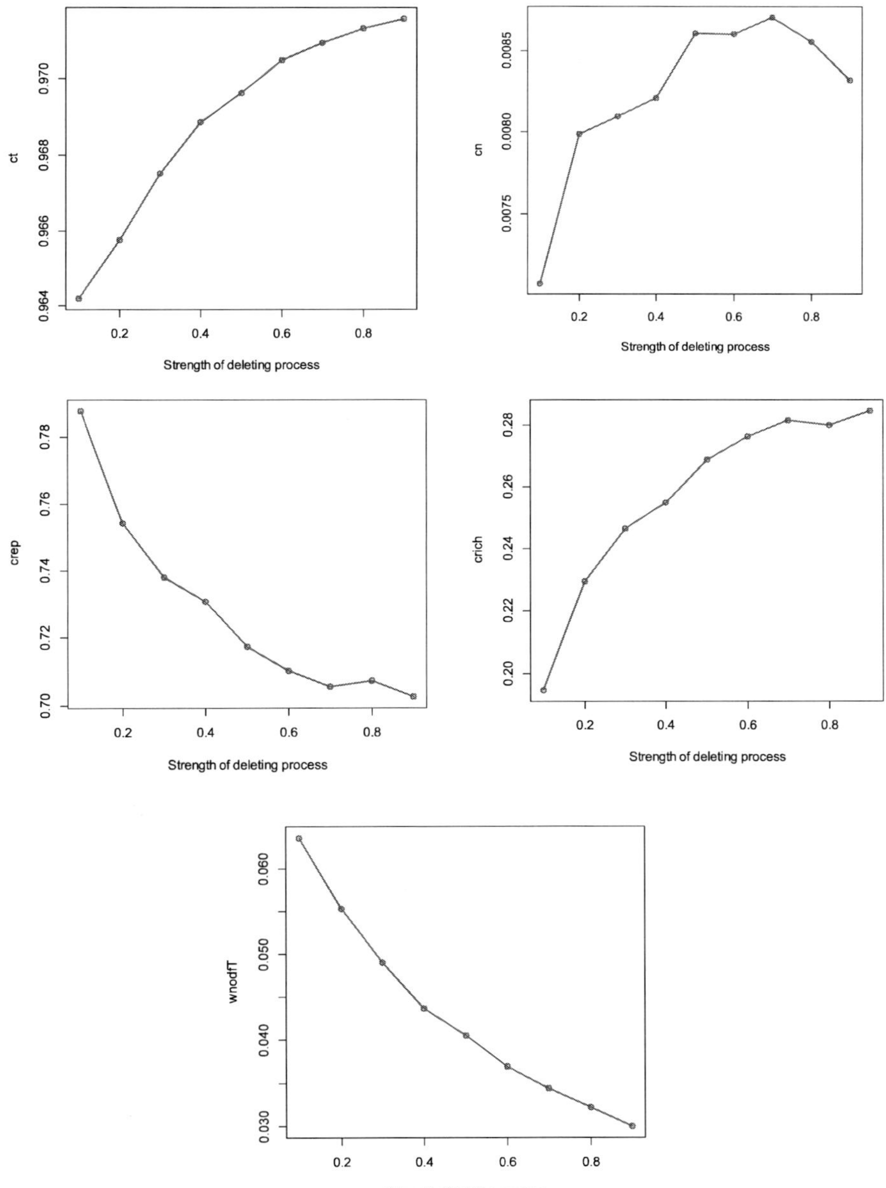

Figure 6. Effect of deleting process on influencing community beta diversity patterns in BCI data sets. Each line indicated the mean from 100 replicates.

In summary, multiple-site nestedness and turnover indices seemed better than other indices, including WNODF, replacement and richness difference, to quantify the effect of species/population changes on influencing community diversity when the community is not in high biodiversity level and when the community is derived from a random perfectly nested matrix (Figures 1-3). However, when the community has a high degree of biodiversity resources and not directly generated from a random perfectly nested matrix (e.g., 100 sites but have over 800 species in BCI tree data set), richness difference and replacement indices seemed better for reflecting species/population change patterns. Thus, I concluded that all the multiples-site indices (except for WNODF index) should be compared and analyzed for ecological communities so as to better reflect community beta diversity variation in a holistic way.

REFERENCES

Almeida-Neto M, Frensel D, Ulrich W (2012) Rethinking the relationship between nestedness and beta diversity: a comment on Baselga (2010). *Glob Ecol Biogeogr* 21:772–777.

Almeida-Neto M, Ulrich W (2011) A straightforward computational approach for measuring nestedness using quantitative matrices. *Environ Model Softw* 26:173–178.

Baselga A (2012) The relationship between species replacement, dissimilarity derived from nestedness, and nestedness. *Glob Ecol Biogeogr* In press.

Baselga A (2010) Partitioning the turnover and nestedness components of beta diversity. *Glob Ecol Biogeogr* 19:134–143. doi: 10.1111/j.1466-8238.2009.00490.x.

Baselga, Jimenez-Valverde A, Niccolini G (2007) A multiple-site similarity measure independent of richness. *Biol Lett* 3:642–645.

Carvalho J, Cardoso P, Borges P, et al. (2012a) Measuring fractions of beta diversity and their relationships to nestedness: a theoretical and empirical comparison of novel approaches. Oikos Doi: 10.1111/j.1600–0706.2012.20980.x. doi: 10.1111/j.1600-0706.2012.20980.x.

Carvalho J, Cardoso P, Gomes P (2012b) Determining the relative roles of species replacement and species richness differences in generating beta-diversity patterns. *Glob Ecol Biogeogr* 21:760–771.

Condit R, Pitman N, Leigh EG, et al. (2002) Beta-diversity in tropical forest trees. *Science* (80-) 295:666–9.

Crowe T, Frost N, Hawkins S (2011) Interactive effects of losing key grazers and ecosystem engineers vary with environmental context. *Mar Ecol Prog Ser* 430:223–234.

Gotelli N, Ulrich W (2010) The empirical Bayes approach as a tool to identify non-random species associations. *Oecologia* 162:463–477.

Larsen T, Williams N, Kremen C (2005) Extinction order and altered community structure rapidly disrupt ecosystem functioning. *Ecol Lett* 8:538–547.

Podani J, Shmera D (2012) A comparative evaluation of pairwise nestedness measures. *Ecography* (Cop). doi: 10.1111/j.1600-0587.2011.07319.x.

Pongsiri M, Roan J, Ezenwa V, et al. (2009) Biodiversity loss affects global disease ecology. *Bioscience* 59:945–954.

Schmera D, Podani J (2011) Comments on separating components of beta diversity. *Community Ecol* 12:153–160.

Ulrich W, Gotelli N (2007) Null model analysis of species nestedness patterns. *Ecology* 88:1824–1831.

Chapter 3

ON THE RELATIONSHIP BETWEEN COMMUNITY-LEVEL BETA DIVERSITY AND PARTITIONING VARIATION FOR SPACE AND ENVIRONMENT: BCI TREE COMMUNITY AS A CASE STUDY

ABSTRACT

Understanding the mechanisms to determine community-level diversity would allow us better quantify the interplay among species, environment and space. In the present study, I quantified the roles of alpha diversity, rarity, beta diversity at community level on influencing partitioning of the variations accounted for space and environment. I proposed two novel multiple-site beta-diversity indices for measuring community-level replacement and richness difference, on the basis of previous work on developing indices for pairwise-site comparison. Then, by combing the newly developed indices with another two multiple-site indices developed previously for measuring community-level turnover and nestendess, I quantified the effects of changing community-level diversity on affecting the partitioning of variation contributed to space and environment.

Keywords: beta diversity, variation partitioning, permanent forest plots, beta diversity partitioning, multiple-site indices

INTRODUCTION

Threats of species extinction are typically driven by a complex set of processes. However, one of the most important factors is the anthropogenic disturbance. Human disturbance on biodiversity extinction has broad implications of biodiversity conservation and ecosystem functioning.

Variation partitioning is a technique widely applied in community ecology (Legendre 2007; Legendre et al. 2005; Peres-Neto et al. 2006). The purpose of variation partitioning is to understand the relative contribution of spatial autocorrelation and environmental filtering on structuring community and better quantify beta-diversity in community ecology (Legendre et al. 2009; Legendre et al. 2005). Variation partitioning has been argued to have the power to capture the variance of compositional patterns of the community, which were thought to drive by spatial autocorrelation and niche deterministic processes (Legendre et al. 2005; Peres-Neto et al. 2006; Smith and Lundholm 2010). However, as I emphasize, the role of human disturbance and other random processes should be also vital to influence the partitioning of the community variation.

Rare species have their unique roles contributing to community diversity (Mi et al. 2012). They are also important to identify conservation priorities (Williams et al. 1996), evaluate extinction risks (He and Hubbell 2011) and so on. However, there is no quantitative assessment on the role of rare species and its abundance on influencing variation partitioning. I might expect that higher rarity degree of species in the community, the variation separated for space should be smaller as the influence of spatial autocorrelation became limited to shorter distances.

Diversity could be separated as three components, alpha diversity, beta diversity and gamma diversity respectively (Whittaker 1972). Beta diversity is the most concerned component in community ecology up to date (Condit et al. 2002; Kraft et al. 2011; Legendre 2007; Legendre et al. 2005).

Recently, biodiversity measurement has been undergoing rapid development, and new indices were proposed for multiple-site similarity (Baselga et al. 2007; Diserud and Ødegaard 2007) and nestendess (Baselga 2010) comparisons.

However, the turnover and nestedness indices (Baselga 2010) for pairwise comparison of sites have been found inconsistent with the expectations (Almeida-Neto et al. 2012; Podani and Shmera 2012; Schmera and Podani 2011). The expectation on the development of multiple-site version of beta diversity indices is desirable (Carvalho et al. 2012b). In light of these, I provided two new metrics to quantify the multiple-site beta-diversity, on the basis of previous work on pair-site indices (Carvalho et al. 2012b), for better capturing the information of species loss.

If human disturbance is said to drive some species and their individuals go extinct at some sites, I might be able to observe a different scenario of variation partitioning because the loss of species and individuals have altered the community structure. Thus, I should quantitatively assess the biodiversity lost caused by human disturbance and its role on changing beta-diversity partitioning. Quantifying the degree of arthropogenic disturbance based on the change of beta-diversity partitioning would have broad implications to better improve the understanding of the interactions between species, humans, environment and space.

Thus, by adopting previously developed and the two novel developed in the present study for multiple-site comparison, I are able to quantify community-level diversity status with some single values. Inspired by these multiple-site measures of diversity, I could quantitatively assess the relationship between community diversity measurements (alpha diversity, turnover and nestedness, replacement and richness difference) and partitioning of variations attributed to space and environment.

In summary, the main objectives of present study is to, 1) develop two multiple-site indices for partitioning beta-diversity into replacement and richness differences; 2) assess the relationships between explained compositional variation partitioned for space and environment and the multiple-site turnover, nestedness, replacement and richness differences; 3) compare the difference of multiple-site turnover, nestedness, replacement and richness differences under different species handling situations.

MATERIALS AND METHODS

Methods for Variation Partitioning

Currently, there are two distinct methods for performing variation partitioning: one is derived from distance matrix, for which I called as distance-based methods. Typically Mantel test (Ruokolainen and Tuomisto 2002; Tuomisto et al. 2012; Tuomisto and Ruokolainen 2006) is the exemplified method for utilizing distance matrix to infer the contributions of space and environment. The other one is derived from the ordination or regression analysis on raw data without calculating distance matrices. For this group of methods I called it as ordination-based methods. Typically, multivariate regression analysis, redundancy analysis and canonical correspondence analysis are the methods widely used techniques, all of which could be put inside this group (Borcard et al. 1992; Legendre et al. 2005; Legendre and Legendre 1998; Peres-Neto et al. 2006). There are some debates on the disadvantages of different methods (Legendre et al. 2008). I are not going to discuss them in detail. Instead, I considered both distinct methods for performing partitioning variations in my study.

Further, I don't try to investigate shared variation which was co-explained by both space and environment, as it could generate negative values (Borcard et al. 1992; Peres-Neto et al. 2006). This phenomenon is hard to explain and understand from ecological perspective.

Assessment the Influence of Community Change on Variation Partitioning

Species Rarity versus Commonness

The rarity of species was calculated on the basis of their abundances and distributional ranges across the 98 sampling plots. I quantified the rarity of species based on the occupancy frequencies and abundance respectively (Gaston 1994; Ricotta et al. 2012). Thus, both density and range rarity were measured.

The percentage of species distributed less than 30% of the total sampling plots was used to measure the rarity of range for the community. In contrast, the percentage of species with abundance less than 30% of the mean abundance of species for the whole community was used to measure the rarity of abundance.

Community Turnover

I used a beta-diversity similarity index accounting for multiple sites (Baselga et al. 2007) to measure the community turnover. The index was formulated as follows,

$$\beta_{tur} = \frac{\sum_{i<j} \min(b_{ij}, b_{ji})}{(\sum_{i} S_i - S_T) + (\sum_{i<j} \min(b_{ij}, b_{ji}))} \quad (1)$$

Where, S_i is the number of species found in site i, while S_T total number of species in the community. b_{ij} and b_{ji} are the number of species exclusive to sites i and j respectively when compared by pairs.

Community Nestedness

I used the nestedness index developed for multiple sites to quantify community nestedness status (Baselga 2010), which is written as,

$$\beta_{nes} = \frac{\sum_{i<j}\max(b_{ij},b_{ji}) - \sum_{i<j}\min(b_{ij},b_{ji})}{2(\sum_i S_i - S_T) + \sum_{i<j}\max(b_{ij},b_{ji}) + \sum_{i<j}\min(b_{ij},b_{ji})} \times \frac{(\sum_i S_i - S_T)}{(\sum_i S_i - S_T) + \sum_{i<j}\min(b_{ij},b_{ji})} \quad (2)$$

The definitions of the letters here are identical to those in community turnover index (1).

Community Replacement

In the present work, I develop novel metrics on the basis of previous work (Carvalho et al. 2012b), which argued that beta-diversity should be partitioned into species replacement and richness difference, so as to generate more ecologically meaningful results. The species replacement index thereby has been developed for pair-site comparison as follows (Carvalho et al. 2012b),

$$\beta_{rep}^{pairs} = 2 \times \frac{\min(b_{ij}, b_{ji})}{a + b_{ij} + b_{ji}} \quad (3)$$

Thus, I now extrapolate this pair-site comparison index into the general form for multiple-site comparison as follows,

$$\beta_{rep} = 2 \times \frac{\sum_{i<j} \min(b_{ij}, b_{ji})}{\left(\sum_i S_i - S_T\right) + \sum_{i<j}(b_{ij} + b_{ji})} \quad (4)$$

Community Richness Difference

Species richness difference is another component of beta-diversity in Carvalho et al.'s work (Carvalho et al. 2012b), which could be written as,

$$\beta_{rich}^{pairs} = \frac{|b_{ij} - b_{ji}|}{a + b_{ij} + b_{ji}} \quad (5)$$

Thus, now I extrapolate this pair-site comparison index into the general form for multiple-site comparison as follows,

$$\beta_{rich} = \frac{\sum_{i<j} \max(b_{ij}, b_{ji}) - \sum_{i<j} \min(b_{ij}, b_{ji})}{2(\sum_i S_i - S_T) + \sum_{i<j} \max(b_{ij}, b_{ji}) + \sum_{i<j} \min(b_{ij}, b_{ji})} \quad (6)$$

Pseudo-Human Disturbance Experiments

For assessing the roles of changing community rarity, beta-diversity and nestedness patterns which were induced by human disturbance, I have two strategies.

The first one is to move the individuals of some species from sites with high abundance to those without the presence of the species but keeping the total abundance of each species unchanged (zero-sum model). This sampling procedure is running with replacement. I choose 30% of species undergoing movement of individuals.

By this manner, homogeneity of the community (between-site similarity) should be enhanced. Thus, this will result into changing rarity, beta-diversity, and nestendess patterns of the community. Thereby, I could observe the corresponding change of variation partitioning.

At the meantime, I also consider another strategy by removing species and/or individuals from the community. I randomly removed some individuals of common species to make them become rare in the community, or randomly removed some species from the calculation. In such a way, I then test the change of variation partitioning according to the changing patterns of the community.

All the analyses were repeated by 500 times for measuring uncertainties of simulations.

Distribution of BCI Trees in 98 Sampling Plots

The distribution of the tree abundances across the BCI sampling plots were derived from (Condit et al. 2002). Originally there were 100 sampling sites, however two of them lacked the environmental data (P40 and P41), thereby being excluded in the analysis. The associated species that were found only in these two plots were excluded in the analysis, resulting a species-site matrix with the dimension of 776×98.

Environmental variables for the analysis included the precipitation and elevation, age of the plot and the geology. Values for the geology variable is categorical, thus I used discrete integers to represent them. Spatial variables included the latitude and longitude of the sampling plots.

RESULTS

Effect of Different Species Handling Methods on Variation Partitioning

In the four species handling procedures, I found that slight alternation of community composition don't remarkably influence the resultant variation partitioning. Thus, variation partition seemed stable when the species handling effects are slight. However, when I have increasing efforts to homogenize or remove/increase species (or individuals) across sites, I found that both the explained variation attributed to space and environment would decrease subsequently.

Effect of Different Species Handling Methods on Community Turnover and Nestedness

I found that in all kinds of species handling processes, community turnover and nestedness indices were totally unaffected. They were quite insensitive to the change of community composition, for either species/population reduction or relocation processes. Thus, I found that the turnover and nestedness indices are really hard to measure different community patterns, congruent with the preceding findings (Carvalho et al. 2012a; Carvalho et al. 2012b).

Effects of Different Species Handling Methods on Community Replacement and Richness Differences

I found that the two novel multiple-site indices developed in the present work could better capture the information of community alternation and compositional change and are tightly associated with the change of variation partitioning.

When manipulated community has decreasing trend of partitioned variation for space and environment, the community richness differences would have an increasing trend, while the community replacement would have a decreasing trend remarkably.

DISCUSSION

Turnover versus Nestedness and Replacement versus Richness Difference

All kinds of species handling procedures could not remarkably change the community nestedness and turnover indices. They tended to fix at some values in all cases of community change. However, I found that each of the species handling procedures could lead to considerable change of replacement and richness difference indices. Thus, replacement and richness difference indices are better to quantify species loss or community alternation in the community, while nestedness and turnover indices are not ecologically informative, consistent with the pervious observations (Almeida-Neto et al. 2012; Carvalho et al. 2012a; Carvalho et al. 2012b; Schmera and Podani 2011).

Thus, it is more appropriate to employ the two novel multiple-site indices (community replacement and richness difference, equations (3) and (4)) to assess community-level beta diversity, so as to generate expected ecological meaningful results (i.e., sensitive to species loss and any kinds of dissimilarity (Carvalho et al. 2012a)).

Change of Rarity during Species Handling Procedures

Deletion of species remarkably changes the community rarity, for either range or density rarity. Rarity of species could contribute to diversity for different levels, e.g., species level, functional level (Dimitriadis and Koutsoubas 2011) and so on.

Comparison of Variation Partitioning Using Mantel Test and CCA

It is found that Mantel test-based variation partitioning is quite sensitive to compositional change. In contrast, CCA-based variation partitioning did not vary a lot when changing composition with different handling methods.

However, through correlation analysis, I found that CCA-based variation partitioning could have higher correlations with multiple-site beta diversity indices when the community is altered. Thus, it showed here CCA ordination methods were better to quantify the fluctuation of beta diversity in the ecological communities.

References

Almeida-Neto M, Frensel D, Ulrich W (2012) Rethinking the relationship between nestedness and beta diversity: a comment on Baselga (2010). *Glob Ecol Biogeogr* 21:772–777.

Baselga A (2010) Partitioning the turnover and nestedness components of beta diversity. *Glob Ecol Biogeogr* 19:134–143. doi: 10.1111/j.1466-8238.2009.00490.x.

Baselga, Jimenez-Valverde A, Niccolini G (2007) A multiple-site similarity measure independent of richness. *Biol Lett* 3:642–645.

Borcard D, Legendre P, Drapeau P (1992) Partialling out the Spatial Component of Ecological Variation. *Ecology* 73:1045.

Carvalho J, Cardoso P, Borges P, et al. (2012a) Measuring fractions of beta diversity and their relationships to nestedness: a theoretical and empirical comparison of novel approaches. Oikos Doi: 10.1111/j.1600–0706.2012.20980.x. doi: 10.1111/j.1600-0706.2012.20980.x.

Carvalho J, Cardoso P, Gomes P (2012b) Determining the relative roles of species replacement and species richness differences in generating beta-diversity patterns. *Glob Ecol Biogeogr* 21:760–771.

Condit R, Pitman N, Leigh EG, et al. (2002) Beta-diversity in tropical forest trees. *Science* (80-) 295:666–9.

Dimitriadis C, Koutsoubas D (2011) Functional diversity and species turnover of benthic invertebrates along a local environmental gradient induced by an aquaculture unit: the contribution of species dispersal ability and rarity. *Hydrobiologia* 670:307–315.

Diserud O, Ødegaard F (2007) A multiple-site similarity measure. *Biol Lett* 3:20–22.

Gaston K (1994) Rarity. Chapman and Hall, London.

He F, Hubbell S (2011) Species–area relationships always overestimate extinction rates from habitat loss. *Nature* 473:368–371.

Kraft NJB, Comita LS, Chase JM, et al. (2011) Disentangling the drivers of β diversity along latitudinal and elevational gradients. *Science* 333:1755–8. doi: 10.1126/science.1208584.

Legendre P (2007) Studying beta diversity: ecological variation partitioning by multiple regression and canonical analysis. *J Plant Ecol* 1:3–8.

Legendre P, Borcard D, Peres-Neto P (2008) Analyzing or explaining beta diversity : Comment. *Ecology* 3238–3244.

Legendre P, Borcard D, Peres-Neto P (2005) Analyzing beta diversity: Partitioning the spatial variation of community composition data. *Ecol Monogr* 75:435–450.

Legendre P, Legendre L (1998) Numerical ecology. *Elsevier Science* BV, Amsterdam.

Legendre P, Mi X, Ren H, et al. (2009) Partitioning beta diversity in a subtropical broad-leaved forest of China. *Ecology* 90:663–674.

Mi X, Swenson NG, Valencia R, et al. (2012) The Contribution of Rare Species to Community Phylogenetic Diversity across a Global Network of Forest Plots. *Am Nat* 180:E17–30. doi: 10.1086/665999.

Peres-Neto P, Legendre P, Dray S, Borcard D (2006) Variation partitioning of species data matrices: estimation and comparison of fractions. *Ecology* 87:2614–2625.

Podani J, Shmera D (2012) A comparative evaluation of pairwise nestedness measures. *Ecography* (Cop). doi: 10.1111/j.1600-0587.2011.07319.x.

Ricotta C, Heathfield D, Godefroid, Mazzoleni S (2012) The effects of habitat filtering on the phylogenetic structure of the urban flora of Brussels (Belgium). *Community Eol* 13:97–101.

Ruokolainen K, Tuomisto H (2002) Beta-diversity in tropical forests. *Science* 297:1439; discussion 1439. doi: 10.1126/science.297.5586.1439a.

Schmera D, Podani J (2011) Comments on separating components of beta diversity. *Community Ecol* 12:153–160.

Smith TW, Lundholm JT (2010) Variation partitioning as a tool to distinguish between niche and neutral processes. *Ecography* (Cop) 33:648–655. doi: 10.1111/j.1600-0587.2009.06105.x.

Tuomisto H, Ruokolainen K (2006) Analyzing or explaining beta diversity? Understanding the targets of different methods of analysis. *Ecology* 87:2697–708.

Tuomisto H, Ruokolainen L, Ruokolainen K (2012) Modelling niche and neutral dynamics: on the ecological interpretation of variation partitioning results. *Ecography (Cop) no–no.* doi: 10.1111/j.1600-0587.2012.07339.x.

Whittaker R (1972) Evolution and measurement of species diversity. *Taxon* 21:213–351.

Williams P, Gibbons D, Margules C, et al. (1996) A comparison of richness hotspots, rarity hotspots, and complementary areas for conserving diversity of British birds. *Conserv Biol* 10:155–174.

Chapter 4

DISTINGUISHING NICHE AND NEUTRAL PROCESSES: ISSUES IN VARIATION PARTITIONING STATISTICAL METHODS AND FURTHER PERSPECTIVES

ABSTRACT

Variance partitioning methods, which are built upon multivariate statistics, have been widely applied in different taxa and habitats in community ecology. Here, I performed a literature review on the development and application of the methods, and then discussed the limitation of available methods and the difficulties involved in sampling schemes. The central goal of the work is then to propose some potential practical methods that might help to overcome different issues of traditional least-square-based regression modeling. A variety of regression models has been considered for comparison. In initial simulations, I identified that generalized additive model (GAM) has the highest accuracy to predict variation components. Therefore, I argued that other advanced regression techniques, including the GAM and related models, could be utilized in variation partitioning for better quantifying the aggregation scenarios of species distribution.

Keywords: multivariate ordination, regression models, general additive models, dispersal limitation, environmental filtering

INTRODUCTION

Variation Decomposition in Community Ecology

It is quite often not only one process regulating and determining community structure. Typically, the combination of multiple processes and their interactions will have profound impacts on resultant community structure. So, it is natural to ask which kinds of processes are dominant, while others are auxiliary. Thus, the variance in response variables can be separated into several parts, and by employing statistical methods, we can identify the contribution and relative importance of different ecological mechanisms.

Figure 1 depicts the methods for performing variation decomposition at different data levels. The methods range from simple linear regression, to multiple regression models, to multivariate regression model, and other multivariate statistical methods.

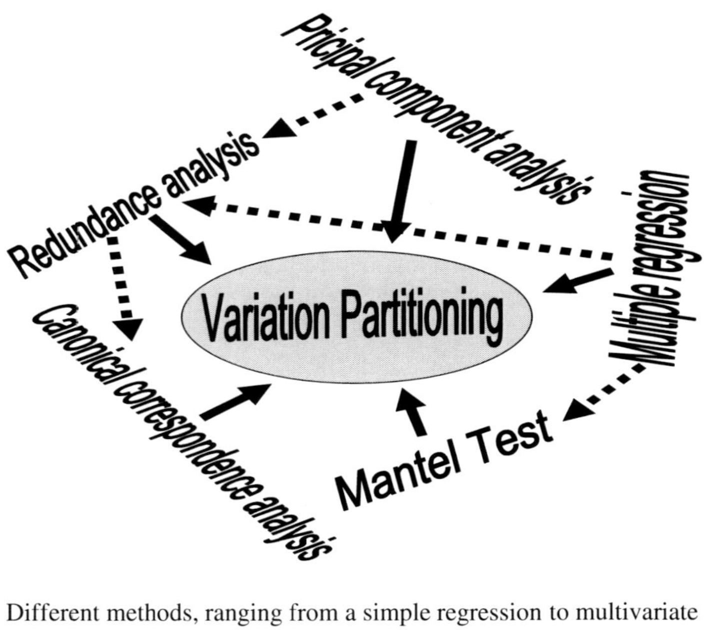

Figure 1. Different methods, ranging from a simple regression to multivariate statistics, have been used for variation partitioning to distinguish environmental and spatial filtering. Solid arrow denotes that the method can be used for performing variation partitioning tests, while dashed arrow denotes that that method (pointed by the head of arrow) can be generalized or deduced from another method.

How Is Variation Partitioning Related to the Debate between Niche and Neutral Processes?

Two mechanisms significantly affect species diversity patterns: neutral and deterministic factors. Environmental descriptors, defining the niche of species, are deterministic; while spatial descriptors, defining the dispersal ability of species, are neutral.

Since Hubbell's neutral theory (Hubbell 2001), a great amount of works tried to predict the power of neutral theory in empirical data. However, most of them failed to support neutral theory (Gotelli and McGill 2006; McGill 2003). There are several ways to test neutral theory. One is to generate individual predictions and test them by regression based on neutral theory. For example, the distance decay of species composition (Gilbert and Lechowicz 2004); the priority effect of juvenile co-occurrence reduction (Gilbert et al. 2008).

Since the repellence of pure neutral theory in empirical test, the mainstream nowadays is to combine both niche and neutrality to explain community structure, and test the relative importance of each component. Therefore, variation partitioning is introduced (Borcard et al. 1992) to detect the contribution from each part of variations. In a work of Tuomisto and Ruokolainen (Tuomisto and Ruokolainen 2006), they suggested that dispersal limitation derived from neutral process can only be tested by using distance-based methods.

How to Perform Variation Partitioning?

To character species composition and environmental variables is a major topic in current ecological research. Redundancy analysis (suited for linear relationships between species composition and environmental variables) and Canonical correspondence analysis (handling nonlinear species-environment relationship) are the two widely used methods to investigate the relationship of environmental variables and species diversity information (Figure 1). Variation partitioning can be used to test and determine the possibilities of individual predictors in influencing species distribution and abundance (Peres-Neto et al. 2006).

Variation partitioning can be divided into four parts: pure environmental variation, pure space variation, mixed environmental and space variation, and unexplained variation (Borcard et al. 1992). Figure 2 showed the schematic map of the variation components which were often complied in previous literature.

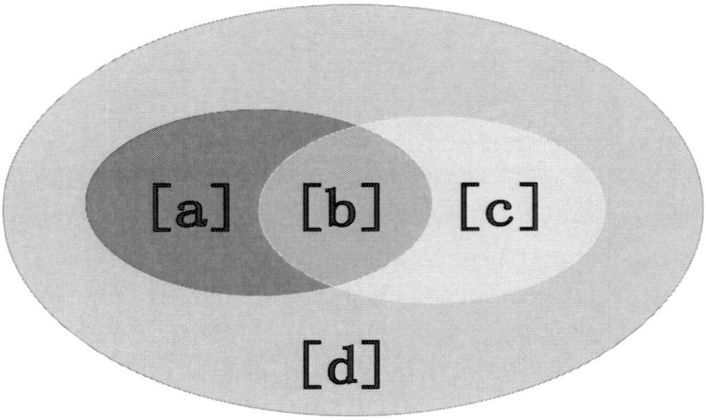

Figure 2. Schematic map showing different components and fractions that are related to variation partitioning. a-pure environmental variation; b-mixed environmental and space variation; c-pure space variation; d-unexplained variation. a+b+c+d=total variance involved in the community data.

In principle, I use the spatial coordinates as the basic spatial descriptors. I can use the eigenvectors derived from the principal components of spatial coordinates, which has been used in some previous works (Dray et al. 2006). Or, I can get the Moran's eigenvector maps (Dray et al. 2006; Sattler et al. 2010), which is a general form of principal coordinates of neighbour matrices (Borcard and Legendre 2002). Both methods use the eigen functions of spatial connectivity matrices, thus they are scale-independent.

When setting spatial descriptors as covariables, I can know the proportion of pure environmental variation. In contrast, when setting environmental variables as covariables, I can deduce the proportion of pure space variation. The mixed environmental and spatial variation can be derived from the subtraction of total known variation-pure environmental variation-pure spatial variation.

It is quite simple to perform comparative studies in ecological data by implementing variation partitioning in regression results. Basically, all kinds of statistical software and tools can implement variation partitioning, as long as they can perform multiple regression analysis. As I stated above, it needs only three times of running regressions, each of which should comprise environmental variables only, spatial variables only and both spatial and environmental variables together. From a convenience perspective, there is a commercial statistical software called Canoco (ter Braak 1986), which is designed for constrained community ordination analysis. Variation partitioning can be implemented using partial CCA method in the package. Besides that, I can perform variation partitioning by using some open-source packages in R software. For example, the command “varpart” can be recalled to perform variation partitioning using “vegan” package (Oksanen et al. 2012).

Literature Searching and Study Classification

To search the available publications relevant to separate the effects of niche processes and neutrality, I used the following databases: Google Scholar, Web of Science, Springlink, Wiley-Blackwell, and Elsevier publishers.

The keywords used for query include “variation partition,” “redundancy analysis,” “niche and neutral processes,” “dispersal limitation and habitat filtering,” “spatial and environmental descriptors.”

Based on the query results, I summarized two basic categories for the subject, which are 1) theoretical and methodological development; and 2) applications on different taxa and habitats. Table 1 summarized all the relevant work on the subject.

From the table, it seemed that most of applications of variation partitioning are on terrestrial ecosystems, and the studied taxa varied from birds to plants. Interestingly, there are only a few relevant works on marine and aquatic ecosystems, with only focus on fish assemblages.

A Review of Current Methods and Possible Problems

There are many possible methods by applying nonlinear regression techniques to reveal the correlation of environmental and spatial variables against species distribution-composition matrix. The possibilities of introducing advanced nonlinear regression model, including local regression methods, general additive models and least partial square methods, can be beneficial to overcome the challenges.

Gilbert and Bennett (Gilbert and Bennett 2010) performed a simulation comparison for analyzing the powers and differences among a variety of variation partitioning methods, most of which are widely used. Typically, the most prevailing methods are the regression on distance matrices (e.g., Mantel test), canonical correspondence analysis and redundancy analysis. Moreover, it is suggested to better retain spatial information by using some kinds of transformation called principal components of neighbor matrices (or Moran's eigenvector maps).

Table 1. A literature review on the methodology and applications of variation partitioning in community ecology

Sub-discipline	Description	Literature
1, Methodological background and development	Advocating or criticizing the variance partitioning and developing relevant tools	1, the original paper describing variation partitioning: Borcard et al. (Borcard et al. 1992) 2, comparison of different methods on performing variation partitioning. For example, Mantel test, multiple regression model, canonical correspondence analysis and so on (Legendre et al. 2008; Legendre et al. 2005) 3, ecological questions that can be addressed by variation partitioning: beta diversity (Legendre et al. 2009; Legendre et al. 2005); neutrality versus niche (Smith and Lundholm 2010; Tuomisto et al. 2012); 4, rebuttal to variation partitioning methods and relevant technical aspects: Tuomisto and Ruokolainen (Tuomisto and Ruokolainen 2006) suggested that distance method (like Mantel test) should be the only choice to test neutral hypothesis. Gilbert and Bennett (Gilbert and Bennett 2010) found that traditional variation partitioning methods have a very restricted power to correctly quantify each part of variances involved in the simulated data. Diniz-Filho et al. (Diniz-Filho et al. 2012) suggested that spatial autocorrelation test (Moran's I index) can be linked to niche and neutrality partitioning.
2, Applications on various taxonomic hierarchy and spatial scales	Applying variation partitioning to different community assemblages across different ecosystems, areas and taxa.	1, terrestrial ecosystems: Oribatid mites (Borcard and Legendre 1994; Lindo and Winchester 2009); Forest birds (Cushman and McGarigal 2002; Pearman 2002); Pteridophyte plants (Jones et al. 2008) 2, marine and aquatic ecosystems: Stream fish community (Steward-Koster et al. 2007); Pelagic fish assemblages (Peltonen et al. 2007)

Despite their wide applications, Gilbert and Bennett (Gilbert and Bennett 2010) found out that all kinds of tools have greatly underestimated each part of variances. For example, they found out that canonical ordination under-fitted the environmental variation, which was simulated in a high amount.

Potential Methods

As found, the under-fit problem of different variation components by traditional ordination methods is largely due to the disability of traditional regression models. This is because all the available methods are built on the basis of least-square estimation of regression coefficients. All the conventional methods have the implicit assumption of homoscedasticity involved in the dataset more or less. Thus, as long as the data were composed of inherent heteroscedasticity, the power of least-square regression was questioned. Simple linear or nonlinear (e.g., polynomial regression and general linear models) fitting will be not possible to remove the impact of shifting data variance in the data set.

Figure 3 showed the impacts of heteroscedasticity are hard to remove when plotting fitted residuals after conventional regression models. Thus, it sounds that a promising method to overcome the under-fit problem identified by Gilbert and Bennett (Gilbert and Bennett 2010) is to adopt advanced regression models. Hence, in the following section of our synthesis is to propose advanced regression tools.

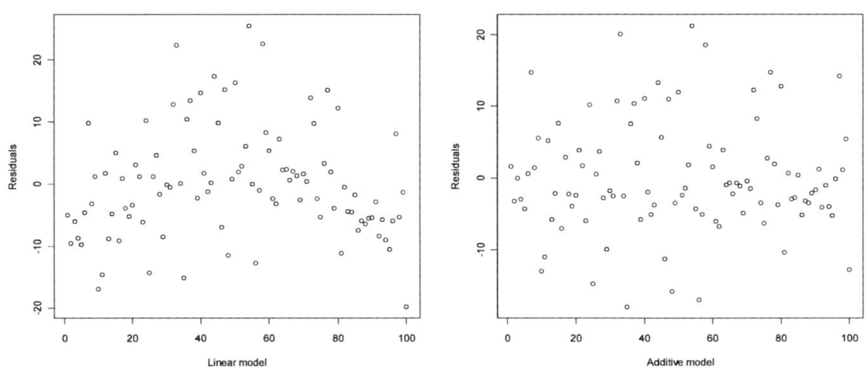

Figure 3. The problem of variance heterogeneity and the effectiveness of additive models compared to linear models to remove variance heterogeneity (left figure is the residuals fitted by simple linear model, while the residuals derived from general additive fitting with Gaussian family is showed in the right figure). It is clear a humped pattern (non-homogenous residuals) still occurs in the left graph.

General Additive Model (GAM) and Relevant Nonlinear Smoothing Methods

The regression models in this category are of course nonlinear, however, another important feature is that they employed completely different ways aiming to solve the problem of heteroscedasticity. General additive model (GAM) typically has the power to remove the problem of variance heterogeneity, with the cost of difficult ecological explanation.

Here I generated a simple relationship between species abundance and one environmental variable with increasing variance across the landscape. Then, I applied different regression models to fit this heterogeneous variance case. The resultant residuals after fitting were as showed in Figure 3, the simple linear model have only R^2-*adjusted*=0.4171, in contrast, GAM returned an R^2-*adjusted*= 0.542. Polynomial regression model is no more than a simple linear model, with R^2-*adjusted*=0.417. Moreover, when checking the regression residuals, I can find out that resultant residuals have no heterogeneity. In contrast, residuals from linear models (the same applied to quadratic nonlinear models, but not showed here) still have variance heterogeneity, and the case becomes worse at both lateral sides of the points.

Implications and Further Perspectives

Spatial Heterogeneity and Distributional Aggregation May Reduce the Power

Typically geographic coordinates are our only choices to measure spatial patterns and drivers of the community structure. However, sometimes it is hard to extract enough spatial information from simple geographic coordinates. In the case of multi-dimensional folding and transformation, Euclidean distances of geographic locations might not be sufficient to capture the variation caused by spatial distances. For example, species distributions typically show the aggregated, rather than random, patterns across different taxa. The driving reasons are usually dispersal-limited colonization and the constraint of habitat heterogeneity, and also biotic interactions, e.g., inter- and intra- specific competition.

Moran's Spatial Scales and Edge Effects

Sampling of different variables at different locations and scales might typically encounter the scale problem. The inconsistence of scales for different variables may lead to bias prediction on disentangling niche and neutrality processes. In such a case, the relative contribution of neutrality and niche processes driving the community structure may be misleading. Figure 4 (upper graph) illustrated the scale issue when doing sampling in fields. Insufficient sampling across the region can give us a rough estimation of spatial gradient, but which is largely departed from the true gradient caused by middle-degree Moran's process. In this case, spatial variation should be overestimated. This issue can happen of course for environmental variables as well.

Edge effects may also inflate the possible separation of environmental filtering and spatial limitation. As showed in the same Figure 4 (lower graph), if the sampling effect is focused on the transitional boundary areas of an environmental variable, the resultant explanation can be that the signaling of environmental filtering is not strong. It is easy to avoid the edge problem for one environmental variable. However, for the case of multiple variables, as it should be hard to predict their transitional boundaries, the sampling scheme can be always coupled with edge effects.

At another side, as known that, both processes can have similar predictions on many facets of community structures. For example, niche process can generate the same distance-decaying pattern as that of neutrality process. In the case like that, typically the resultant community structure is co-dominated by both mechanisms and hard to separate without additional information about the community. Thus, it might be not effective to use partial regression techniques to separate niche and neutrality processes.

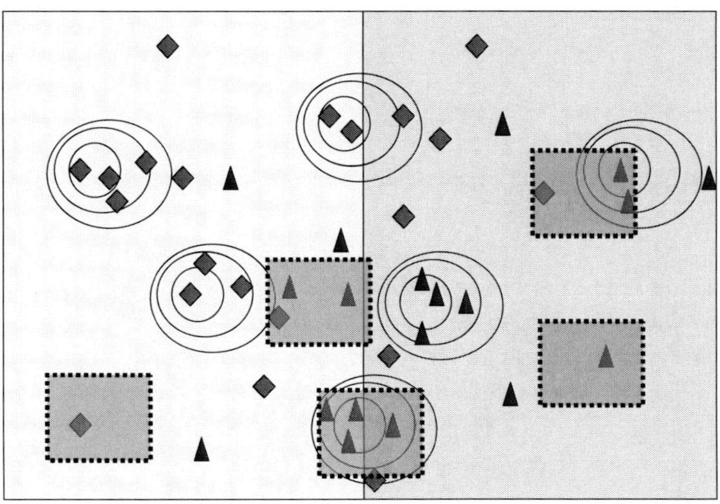

Figure 4. Continued.

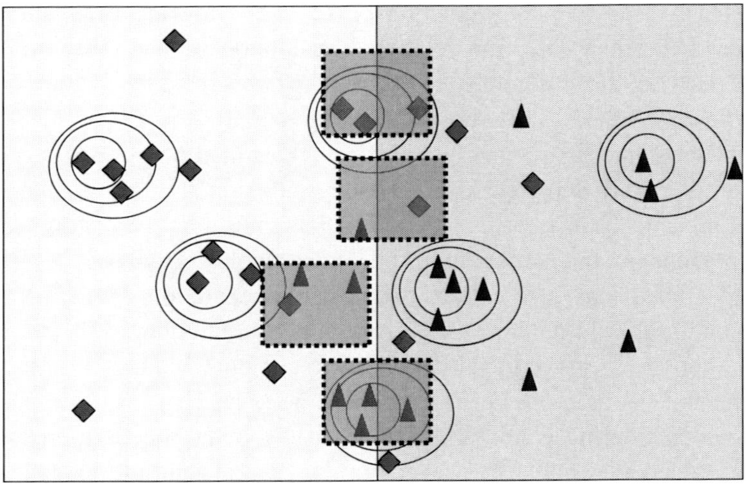

Figure 4. Sampling biases caused by Moran's spatial scales and edge effects. Two rectangular areas indicate two distinct levels of an environmental variable (e.g., precipitation, elevation, temperature and so on). Square and triangle points with red and blue colors indicated two species. Ellipse circles represent density of species population. Transparent gray squares represent sampling plots across the region. In principle, the community bounded by the large square is structured by environmental filtering. The upper graph illustrates the insufficient sampling case, which make the wrong conclusion that environmental filtering is not important to capture the beta-diversity. The lower graph illustrate the edge sampling case, which make the wrong conclusion that spatial filtering is much more important than environmental filtering to structure the community.

Variable Selection Process

Maybe variable selection is an improved way to better capture the most correlated variation information for spatial and environmental drivers. In tradition, backward or forward variable selection procedure is applied to choose the optimal subsets of variables. In the case of variation partitioning, I can do the variable selection separately, then obtain the most correlated spatial and environmental predictors to perform variation decomposition. This method did provide the most significant variables, but a small change of variable subset will lead to a great amount of changes in the resulting variation. Thus, the discrete process may reduce the prediction accuracy when new variables are included (Tibshirani 1996). Fortunately, I have other advanced model selection methods, for example, Lasso and ridge regressions can be good alternatives for choosing good candidates in variation partitioning.

An Integrative Way to Partition and Understand Ecological Communities

The explanation is the most challenging thing for general additive models, although it has a higher appealing prediction power than linear models. The comprising manner is to use simple linear or nonlinear regression to fit the data when obtaining the possible curve pattern inspired by GAM models. Then, combining the fitted adjusted R^2 from GAM and fitted line from either linear or simple nonlinear regression models, I can clearly and easily explain the possible relationships and mechanisms that might dominate the community data.

REFERENCES

Borcard D, Legendre P (2002) All-scale spatial analysis of ecological data by means of principal coordinates of neighbor matrices. *Ecol Modell* 153:51–68.

Borcard D, Legendre P (1994) Environmental control and spatial structure in ecological communities: an example using oribatid mites (Acari, Oribatei). *Environ Ecol Stat* 1:37–61.

Borcard D, Legendre P, Drapeau P (1992) Partialling out the Spatial Component of Ecological Variation. *Ecology* 73:1045.

ter Braak CJF (1986) Canonical correspondence analysis : a new eigenvector technique for multivariate direct gradient analysis. *Ecolog*y 67:1167–1179.

Cushman S, McGarigal K (2002) Hierarchical, multi-scale decomposition ofspecies-environment relationships. *Landsc Ecol* 17:637–646.

Diniz-Filho J, Siqueira T, Padial A, et al. (2012) Spatial atucorrelation analysis allows disentangling the balance between neutral and niche processes in metacommunities. *Oikos* 121:201–210.

Dray S, Legendre P, Peres-Neto PR (2006) Spatial modelling: a comprehensive framework for principal coordinate analysis of neighbour matrices (PCNM). *Ecol Modell* 196:483–493.

Gilbert B, Bennett JR (2010) Partitioning variation in ecological communities: do the numbers add up? *J Appl Ecol* 47:1071–1082. doi: 10.1111/j.1365-2664.2010.01861.x.

Gilbert B, Lechowicz M (2004) Neutrality, niches, and dispersal in a temperate forest understory. *Proc Natl Acad Sci* 101:7651–7656.

Gilbert B, Srivastava DS, Kirby KR (2008) Niche partitioning at multiple scales facilitates coexistence among mosquito larvae. *Oikos* 117:944–950.

Gotelli N, McGill B (2006) Null versus neutral models: what's the difference? *Ecography* (Cop) 29:793–800.

Hubbell SP (2001) The Unified Neutral Theory of Biodiversity and Biogeography (MPB-32) (Monographs in Population Biology). Princeton University Press.

Jones M, Tuomisto H, Borcard D, et al. (2008) Explaining variation in tropical plant community composition: influence of environmental and spatial data quality. *Oecologica* 155:593–604.

Legendre P, Borcard D, Peres-Neto P (2008) Analyzing or explaining beta diversity : Comment. *Ecology 3*238–3244.

Legendre P, Borcard D, Peres-Neto P (2005) Analyzing beta diversity: Partitioning the spatial variation of community composition data. *Ecol Monogr* 75:435–450.

Legendre P, Mi X, Ren H, et al. (2009) Partitioning beta diversity in a subtropical broad-leaved forest of China. *Ecology* 90:663–674.

Lindo Z, Winchester NN (2009) Spatial and environmental factors contributing to patterns in arboreal and terrestrial oribatid mite diversity across spatial scales. *Oecologia* 160:817–25.

McGill B (2003) A test of the unified neutral theory of biodiversity. *Nature* 422:881–885.

Oksanen J, Blanchet F, Kindt R, et al. (2012) vegan: Community Ecology Package. R package version 2.0-4. http://CRAN.R-project.org/package =vegan.

Pearman P (2002) The scale of community structure: habitat variation and avian guids in tropical forest understory. *Ecol Monogr* 72:19–39.

Peltonen H, Luoto M, Pakkonen J, et al. (2007) Pelagic fish abundance in relation to regional environmental variation in the Gulf of Finland, northern Baltic Sea. *ICES J Mar Sci* 64:487–495.

Peres-Neto P, Legendre P, Dray S, Borcard D (2006) Variation partitioning of species data matrices: estimation and comparison of fractions. *Ecology* 87:2614–2625.

Sattler T, Borcard D, Arlettaz R, et al. (2010) Spider, bee, and bird communities in cities are shaped by environmental control and high stochasticity. *Ecology* 91:3343–3353.

Smith TW, Lundholm JT (2010) Variation partitioning as a tool to distinguish between niche and neutral processes. *Ecography* (Cop) 33:648–655. doi: 10.1111/j.1600-0587. 2009.06105.x.

Steward-Koster B, Kennard M, Harch B, et al. (2007) Partitioning the variation in stream fish assemblages within a spatio-temporal hierarchy. *Mar Freshw Res* 58:675–686.

Tibshirani R (1996) Regression shrinkage and selection via the Lasso. *J R Stat Soc* B 58:267–288.

Tuomisto H, Ruokolainen K (2006) Analyzing or explaining beta diversity? Understanding the targets of different methods of analysis. *Ecology* 87:2697–708.

Tuomisto H, Ruokolainen L, Ruokolainen K (2012) Modelling niche and neutral dynamics: on the ecological interpretation of variation partitioning results. *Ecography* (Cop) no–no. doi: 10.1111/j.1600-0587.2012.07339.x.

Chapter 5

ESTIMATING ANCESTRAL RANGES IN HISTORICAL BIOGEOGRAPHY: ON THE DEVELOPMENT OF ALTERNATIVE RANGE EVOLUTION MODELS FOR MAXIMUM LIKELIHOOD INFERENCE

ABSTRACT

In this report, we developed and compared several range evolution transition models specifically for reconstructing ancestral ranges of species' distribution using maximum likelihood method. Comparison of these alternative range evolution methods showed a high degree of consistence but some slight differences still existed. The binary range presence/absence (1/0) transition model, with the assumption of range independence, seemed the best among the proposed four models when reconstructing the ancestral ranges for the Hawaiian *Psychotria* plants. Conclusively, these range evolution models should have potential applications in biogeographic analysis for inferring species' ancestral distribution.

Keywords: matrix exponential, maximum likelihood, biogeographic analysis, ancestral state reconstruction

INTRODUCTION

Ancestral range reconstruction (Bremer 1992), is one of central topics in historical biogeography, has been growingly applied in biogeographic analysis of taxas across different areas and habitats. There are a couple of methods dealing with ancestral range reconstruction. For example, weighted area analysis (Hausdorf 1998); dispersal-vicariance analysis (DIVA: (Ronquist 1997)) and modified extensions (e.g., (Nylander et al. 2008; Harris and Xiang 2009; Yu et al. 2010)); dispersal-extinction-cladogenesis likelihood method (Ree and Smith 2008). More recently, the method of trait-speciation diversification (Maddison et al. 2007) has been adopted into geographic range evolution study as well (e.g., (Goldberg et al. 2011)).

Previous works (Donoghue and Moore 2003; Lamm and Redelings 2009) summarized and criticized the available ancestral range reconstruction methods, and in principle, they can be classified into several types: generalized parsimony methods (e.g., (Ronquist 1994, 1997)), likelihood-based methods (e.g., (Ree et al. 2005; Ree and Smith 2008)); and recently Bayesian probabilistic methods (e.g., (Nylander et al. 2008; Sanmartin et al. 2008)).

Likelihood approaches have been widely applied in phylogenetics. Its capability to identify the correct tree compared to parsimony methods thus received growing recognition and application (Felsenstein 2004). Also, compared to DIVA method, likelihood method can handle more realistic speciation cases and area heritance including allopatric, sympatric and peripatric speciations (Ree et al. 2005). In that work, the Monte Carlo method was used and the computational speed was extremely low and not possible to handle a large number of operating area units. Subsequently, a better version called DEC likelihood method using matrix exponential technique has been released (Ree and Smith 2008).

One interesting question for maximum likelihood estimation of ancestral ranges should be the evolutionary models built specifically for ranges of species. Although there were a lot of rate transition models, most of them were developed specifically under the context of molecular or trait evolution, which might not be suitable for distributional range evolution. Species distribution across different geographic ranges actually might have different aspects compared to the evolution of morphological traits or DNA sequences. For example, dispersal across biogeographic boundaries could be extremely more frequent, or rarer than expected. Ranges without the presence of focused species should be expected to have low or no speciation/mutation rates to transfer from state 0 to 1, etc.

Therefore, in the present study we developed and assessed several possible range evolution transition models, which might have potential implications of applications in phylogeographic studies and biogeographic inferences of ancestral ranges.

Materials and Methods

Range Evolution Models

Range Evolution Model 1 (M1)

The standard character evolution model assumes site independence. The model is simply a binary state transition matrix but the explanation of the state transitions was biogeography-based. Thus, this matrix is specific for a given distributional site; while 0 denotes the absence of the distributional range while 1 presence of the range for the species. The transition matrix Q was written as follows,

$$Q = \begin{matrix} \\ 0 \\ 1 \end{matrix} \begin{pmatrix} 0 & 1 \\ -d & d \\ e & -e \end{pmatrix}$$

Range evolution model 2 (M2):

In this model, all the sites for species distribution are considered simultaneously, therefore, the state of zero (absence of species) could be possible to convert into presence state due to extreme dispersal events. s is a scaling factor to indicate extreme dispersal events. Here we set s=0.0001. Here, 0 and 1 denote the absence and presence of the ancestral range in the reconstruction process. This model is site-dependent, thus we actually put all the sites in the same matrix, and the first two columns denote the absence (0) or presence (1) status of the range 1, while the second two columns absence (0) or presence (1) status of range 2…. The transition matrix Q was written in a form as follows,

$$
Q=\begin{array}{c|cccccccc}
 & 0 & 1 & 0 & 1 & 0 & 1 & 0 & 1 \\
\hline
0 & -sd & sd & & & & & & \\
1 & e & -e-3d & & d & & d & & d \\
0 & & & -sd & sd & & & & \\
1 & & d & e & -e-3d & & d & & d \\
0 & & & & & -sd & sd & & \\
1 & & d & & d & & -e-3d & & d \\
0 & & & & & & & -sd & sd \\
1 & & d & & d & & d & & -e-3d
\end{array}
$$

Range evolution model 3 (M3):

In this model, all sites could be allowed to have connectivity. Thus, all species could disperse from one site to another side with mean dispersal rate. Compared to the previous works (Sanmartin et al. 2008; Sanmartín et al. 2010), carrying capability of the sites was not considered (we performed a comparative study as well by considering carrying capability of the sites as discussed later). Also, the advantage of our range evolution model could extend the rate matrix with any sizes of ranges. In contrast, the general time-reversible (GTR) molecular character model employed in (Sanmartin et al. 2008; Sanmartín et al. 2010) could only handle four distributional ranges, which was simply derived from DNA base-pair evolution (ATGC). Here, 0 denotes zero range (no existence of the range), 1, 2, 3… denotes each of all the single ranges in the model.

$$
Q=\begin{array}{c|cccccccc}
 & 0 & 1 & 2 & 3 & 4 & 5 & 6 & 7 \\
\hline
0 & 0 & 0 & 0 & 0 & 0 & 0 & 0 & 0 \\
1 & e & -e-6d & d & d & d & d & d & d \\
2 & e & d & -e-6d & d & d & d & d & d \\
3 & e & d & d & -e-6d & d & d & d & d \\
4 & e & d & d & d & -e-6d & d & d & d \\
5 & e & d & d & d & d & -e-6d & d & d \\
6 & e & d & d & d & d & d & -e-6d & d \\
7 & e & d & d & d & d & d & d & -e-6d
\end{array}
$$

Range evolution model 4 (M4):

This model is similar to the range evolution matrix in the DEC method, which was developed in (Ree and Smith 2008). The codes for each range were identical: 0 denotes zero range, 1, 2, 3 denotes a single range in the model, while 12, 13... were the combined range from some of the single ranges. The slight difference of our model with the DEC transition matrix is that the dispersal rates from combined ranges to combined ranges (for example dispersal from range 12 to 123) were set to the same values as those from single ranges to combined ranges (for example dispersal from range 12 to 123) as follows,

$$Q = \begin{array}{c} \\ 0 \\ 1 \\ 2 \\ 3 \\ 12 \\ 13 \\ 23 \\ 123 \end{array} \begin{pmatrix} 0 & 1 & 2 & 3 & 12 & 13 & 23 & 123 \\ & & & & & & & \\ e & & & & d & d & & \\ e & & & & d & & d & \\ e & & & & & d & d & \\ & e & e & & & & & d \\ & e & & e & & & & d \\ & & e & e & & & & d \\ & & & & e & e & e & \end{pmatrix}$$

Maximum Likelihood Estimation

For simplicity here, for maximum likelihood estimation, we assumed that dispersal rate and extinction rate has only one value across different ranges (as showed in the Q matrices above). We firstly estimated the dispersal and extinction rates and the ancestral range at the root, and then used the candidate parameters to infer the ancestral ranges at each internal node.

In a given branch with time length t, the transition probability is written as,

$$P(t) = e^{Qt}$$

This form was applied to all the branches and then summed together to obtain the resultant likelihood of the inference of the ancestral states and dispersal/extinction rates.

An Example

Hawaiian Psychotria

The example was derived from the data set used in Ree and Smith (Ree and Smith 2008) and Nepokroeff et al. (Nepokroeff et al. 2003), which included 17 Hawaiian understory tree genus *Psychotria* species (Rubiaceae).

Results

Comparison of Ancestral Range Inference from Different Range Evolutionary Models

In principle, we found that the ancestor states at the root and internal nodes identified by the maximum parsimony method (Nepokroeff et al. 2003), the DEC likelihood method (Ree and Smith 2008) and the four range evolution models proposed in our study using standard maximum likelihood inference were highly similar. Table 1 showed the ancestral states for each of the internal nodes derived from the above 4 range evolution models with relevant maximum likelihood values. We could find out that for most of the internal nodes, the reconstructed ancestral ranges were basically identical.

The Model 4 estimated ancestral states for both the nodes x[7] and x[12] as Maui Nui, which was different from the reconstruction results present in (Ree and Smith 2008) using Lagrange program. Possibly, it is most likely that because the cladogenensis process has not been included in our maximum likelihood framework.

Table 1. Maximum likelihood estimation of dispersal and extinction rates of different range evolution models and the ancestral range states reconstruction. Codes for the areas: 1, Kaua'i, 2, O'ahu, 3, Maui Nui and 4: Hawai'i. Node x[16] was in bold fonts, representing the root of the phylogenetic tree

Models		M1	M2	M3	M4
Range evolution parameters	Dispersal rate	1	1	1	1
	Extinction rate	1	1	0	1
Log-likelihood		-56.025	-18.509	-23.173	-49.806
		Reconstructed ancestral ranges			
Internal nodes	x[1]	3	3	3	3
	x[2]	2	2	2	23
	x[3]	2	2	2	2
	x[4]	3	3	3	3
	x[5]	3	3	3	3
	x[6]	3	3	3	3
	x[7]	1	1	3	3
	x[8]	1	1	1	1
	x[9]	1	4	1	1
	x[10]	34	3	3	34
	x[11]	2	2	2	3
	x[12]	1	1	1	4
	x[13]	1	1	1	1
	x[14]	1	4	1	1
	x[15]	1	1	1	1
	x[16]	**1**	**4**	**1**	**1**
	x[17]	1	1	1	1

Changes of Transition Model Parameters on Ancestral State Fitting

We varied the rates of dispersal and extinction differently to quantitatively understand the role of the two parameters on influencing ancestral state reconstruction. Resulting fitting outcomes were present in Table 2. As clearly showed, when dispersal rate was increased, while extinction rate fixed, the ancestral states have more probabilities in forms of combined areas with two or more islands. In contrast, while varying extinction rate and fixing dispersal rate, we could find that the ancestral states typically in forms of a single island.

DISCUSSION

Comparison of Different Range Evolution Models

Based on the results, the standard binary range transition model (Model 1) outperformed than any other models, which was highly consistent with the previous work (Nepokroeff et al. 2003). Potential problems of the model is that it always assumed that different ranges have independent transition rates between states 0 and 1, which seemed apparently not real as dispersal among the ranges should be very possible. However, site-independence assumption was quite successful because when we checked the situations presented in the Model 2, in which we put all sites together to form a large Q matrix, and the direct connectivity among the sites had been assumed. In this model the possible site-dependence situations were taken into account. However, the reconstruction results showed some unexpected predictions. The ancestral range at the root (4-Hawai'i) was totally different from any other rate models and previous work (Nepokroeff et al. 2003; Ree and Smith 2008). Thus, consideration of site-dependence made this model being unusual at predicting ancestral ranges at the root. This result thus may suggest that simple binary range transition model presented in model 1 was robust to infer ancestral ranges.

Range evolution model 3 is similar to but further extends the generalised time-reversible (GTR, (Tavaré 1986)) model, which was designed for molecular evolution. The GTR models have been employed to infer ancestral ranges of plants (Sanmartin et al. 2008; Sanmartín et al. 2010). However, these preceding studies could only limit the island/range number up to 4, for the purpose to fit the size of GTR model. In contrast, the M3 developed in the present work could greatly expand the capability of handling multiple ranges. Of course, our M3 model did not consider the carrying capability of ranges (Sanmartin et al. 2008; Sanmartín et al. 2010), but could be easily incorporated by adding additional terms in the transition matrix.

Finally, model 4 was similar to the one developed by (Ree and Smith 2008). From the comparison among the models, we found that this model tended to reconstruct combined ancestral ranges (i.e., 23, 34) for internal nodes (for internal nodes 2 and 10). One possible reason is the weights given to the dispersal rates in those combined ranges, leading to the ancestral states tending to be combined ranges. Thus, this phenomenon might be a reason why the cladogenesis process has to be integrated for reducing the increasing tendency of merging ranges in the internal nodes.

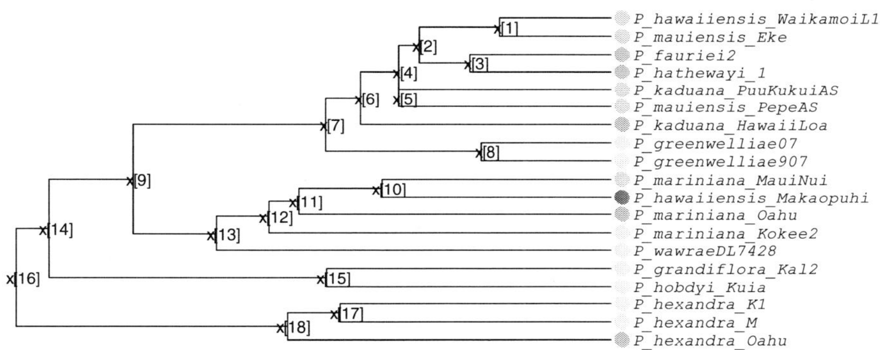

Figure 1. Assignment of codes for internal nodes of the Hawaiian *Psychotria* species.

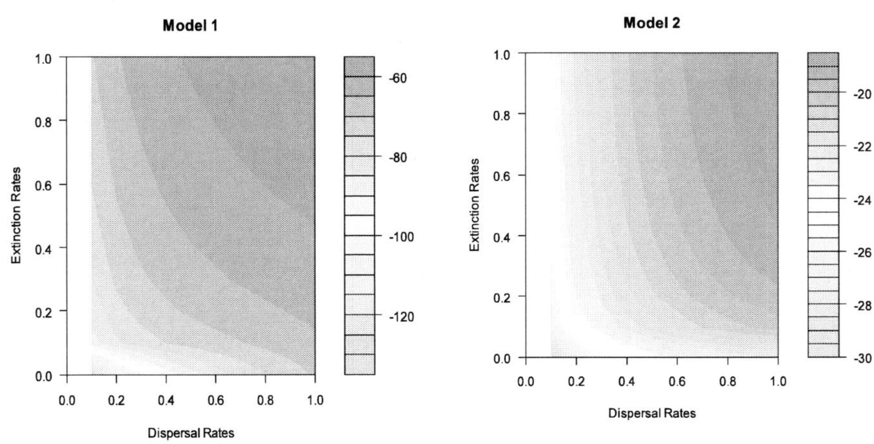

Figure 2. Continued.

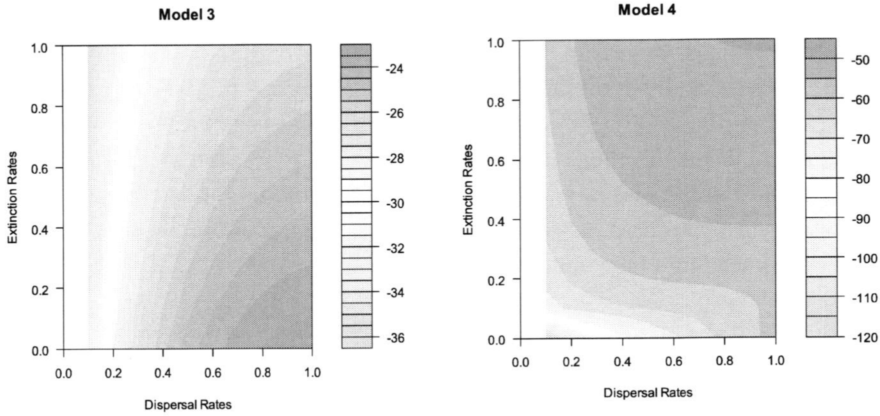

Figure 2. Likelihood contour plot for alternative range evolution models with different dispersal and extinction rates.

On the Estimation of Dispersal and Extinction Rates

Interestingly, we found that it always tended to touch the value boundary [0~1] for the dispersal and extinction rate estimation in a global maximum likelihood process. Contour plot of likelihood for different combinations of dispersal and extinction rates further verified this phenomenon (Figure 2). Thus, the possible reason attributed for that might be due to the independent relationship between the extinction and dispersal rates among the ranges. This independence thus made both parameters tend to be as large as possible (except for the case of Model 3) to maximize the global likelihood at the root.

FURTHER PERSPECTIVES

Developing suitable evolutionary models for inferring phylogenetic history of species historical distributional ranges should be necessary for better quantifying historical biogeographic patterns of species. Our present study showed that different evolutionary model showed slight discrepancies in estimating ancestral ranges for Hawaiian *Psychotria* species. The high congruence of course should indicate that the historical colonization and invasion pathways of this taxonomic group in the four islands should be quite informative. Further tests on the effectiveness of these alternative evolutionary models should be applied to different taxonomic groups at different spatial scales and locations, for the purpose of better unraveling historical biogeographic patterns of species.

REFERENCES

Bremer K (1992) Ancestral areas: a cladistic reinterpretation of centre of origin concept. *Syst. Biol.* 41:436–445.

Donoghue M, Moore B (2003) Toward an integrative historical biogeography. *Integr. Comp. Biol.* 43:261–270.

Felsenstein J (2004) Inferring phylogenies. Sinauer Associates, Sunderland MA.

Goldberg E, Lancaster L, Ree R (2011) Phylogenetic inference of reciprocal effects between geographic range evolution and diversification. *Syst. Biol.* 60:451–565.

Harris A, Xiang Q (2009) Estimating ancestral distributions of lineages with uncertain sister groups: a statistical approach to dispersal-vicariance analysis and a case using Aesculus L. (Sapindaceae) including fossils. *J. Syst. Evol.* 47:349–368.

Hausdorf B (1998) Weighted ancestral area analysis and solution of the redundant distribution problem. *Syst. Biol.* 47:445–456.

Lamm K, Redelings B (2009) Reconstructing ancestral ranges in historical biogeography: properties and prospects. *J. Syst. Evol.* 47:369–382.

Maddison W, Midford P, Otto S (2007) Estimating a binary character's effect on speciation and extinction. *Syst. Biol.* 56:701–710.

Nepokroeff M, Sytsma K, Wagner W, Zimmer E (2003) Reconstructing ancestral patterns of colonization and dispersal in the Hawaiian understory tree genus *Psychotria* (Rubiaceae): A comparison of parsimony and likelihood approaches. *2Systematic Biol.* 52:820–838.

Nylander J, Olsson U, Alstrom P, Sanmartin I (2008) Accounting for phylogenetic uncertainty in biogeography: a bayesian approach to dispersal-vicariance anaysis of the Thurshes (Aves: Turdus). *Syst. Biol.* 57:257–268.

Ree R, Moore B, Webb C, Donoghue M (2005) A likelihood framework for inferring the evolution of geogrpahic range on phylogenetic trees. Evolution (N. Y). 59:2299–2311.

Ree R, Smith S (2008) Maximum likelihood inference of geographic range evolution by dispersal, local extinction and cladogenesis. *Syst. Biol.* 57:4–14.

Ronquist F (1994) Ancestral Areas and Parsimony. *Syst. Biol.* 43:267–274.

Ronquist F (1997) Dispersal-Vicariance Analysis: A New Approach to the Quantification of Historical Biogeography. *Syst. Biol.* 46:195–203.

Sanmartín I, Anderson CL, Alarcon M, Ronquist F, Aldasoro JJ (2010) Bayesian island biogeography in a continental setting: the Rand Flora case. *Biol. Lett.* 6:703–7.

Sanmartin I, Mark P Van der, Ronquist F (2008) Inferring dispersal: a Bayesian approach to phylogeny-based island biogeogarphy, with special reference to the Canary Islands. *J. Biogeogr.* 35:438–449.

Tavaré S (1986) Some Probabilistic and Statistical Problems in the Analysis of DNA Sequences. *Lect. Math. Life Sci. American Math. Soc.* 17:57–86.

Yu Y, Harris A, He X (2010) S-DIVA (statistical dispersal-vicariance analysis): a tool for inferring biogeographic histories. *Mol. Phylogenet. Evol.* 56:848–850.

Chapter 6

ON THE RELATIONSHIPS OF CLADE AGE, SPECIES RICHNESS, AND PHYLOGENETIC DIVERSITY OF GLOBAL MAMMALS

ABSTRACT

In the present report, the relationships between clade age, species richness, and phylogenetic diversity were evaluated for global mammal phylogeny by using non-phylogenetic and phylogenetic regression methods. The results showed that, there are significantly positive relationships between clade age, species richness and phylogenetic diversity for global mammals, regardless of alternative regression methods used. Further, diversification rates of mammals were best estimated to be constant over the evolutionary time, implying there were no rate-shifting patterns for the tempo of species diversification. As such, diversification rates are impotent to predict species richness and/or phylogenetic diversity. In conclusion, clade age is tightly associated with and predicts species richness and diversity for global mammals.

Keywords: clade history, macroevolution, species diversification, macroecology, phylogenetic comparative methods

INTRODUCTION

The relationship between clade age and species richness is one of recent interests in macroevolutionary studies (Rabosky and Adams 2012; Rabosky et al. 2012). Clade age has been thought to relate to species richness because of the fact that older clades could have more time to diversify (McPeek and Brown 2007; Etienne et al. 2012). As such, older age the clade is, higher richness it should have, resulting into a positive clade age-species richness relationship. However, whether such a relationship is universal is still controversial. Some previous studies showed that clade age could predict species richness (McPeek and Brown 2007; Etienne et al. 2012), while others argued that there was no a clear relationship between clade age and species diversity (Rabosky et al. 2012).

Global mammalian phylogeny has been well established (Bininda-Emonds et al. 2007) and the tree file was widely used by many subsequent studies (Cooper et al. 2011; Davies and Buckley 2012). However, it is still largely unknown what kinds of biological processes are driving the contemporary diversity patterns of global mammals. Although there are some previous studies have unraveled the relative importance of ecological and historical factors on structuring the distribution of global mammals (Grenyer et al. 2006; Belmaker and Jetz 2012; Jetz and Fine 2012) or the global phylogenetic diversity patterns of mammals (Safi et al. 2011; Davies and Buckley 2012), few studies focused on the linkage between global mammalian richness and associated evolutionary history. Moreover, there indeed existed a previous study (Rabosky et al. 2012) working on the relationship between clade age and species richness, but it focused on all eukaryotic taxa, without any specific focuses on mammals.

In a summary, for the present report, the central aim is to quantify the relationship between phylogenetic clade age and species richness over the evolutionary history for worldwide mammals.

MATERIALS AND METHODS

Global Mammal Phylogeny

Global mammalian phylogeny was derived from a previous study (Bininda-Emonds et al. 2007). This tree was used for all the subsequent analyses. To infer the relationship between clade age and species diversity, it is necessary to extract subclades from the whole phylogeny. As such, the evolutionary age for each subclade was calculated as the phylogenetic distance between the root and the internal node leading to the focused subclade. The corresponding species richness is defined as the external tips for the focused subclade, while the corresponding phylogenetic diversity is calculated as the sum of all branch lengths for the subclade.

To reveal the possible relationships between clade age and species richness and/or phylogenetic diversity, I performed both non-phylogenetically ordinary least-square regression analysis (OLS) and phylogenetical general regression analysis (PGLS). As different subclades are not independent from each other, it is necessary to consider phylogenetic adjustment by performing PGLS.

Tempo of Species Diversification

I considered implementing different diversification rate-shifting models to fit the worldwide mammalian phylogeny, as proposed in previous studies (Rabosky 2006a; Rabosky and Lovette 2008a; b). In specific, three methods implemented by an R package "laser" (Rabosky 2006b) were considered for comparison, including constant-speciation and constant-extinction model (CONSTANT), decreasing-speciation and constant -extinction model (SPVAR), and constant-speciation and increasing-extinction model (EXVAR). These models have been used to test the temporal shifting pattern of diversification rates of different taxa (Rabosky and Lovette 2008a; b).

In all the models, there are four parameters needed for estimation (Rabosky and Lovette 2008b): constant initial speciation rate λ_0, constant initial extinction rate μ_0, parameter k is for characterizing time-dependent shifting pattern of speciation rate, and parameter z is for characterizing time-dependent shifting pattern of extinction rate. For CONSTANT model, only λ_0 and μ_0 are required to estimate; for SPVAR model, λ_0, μ_0 and k are estimated; for EXVAR model, λ_0, μ_0 and z are estimated.

RESULTS

It was found that both species richness and phylogenetic diversity are significantly correlated with clade age throughout the evolutionary time, regardless of the regression methods used (Figures 1 and 2; Table 1).

Moreover, species richness and phylogenetic diversity are strongly correlated (Figure 3). As such, clade age could predict species richness and phylogenetic diversity pretty well. Log-transformation of both clade age and richness/diversity didn't change the present quantitative results.

Comparative diversification models suggested that diversification rates of species over evolutionary time followed a constant pattern as the CONSTANT model was selected as the best model with lowest AIC value (Table 2).

Table 1. Statistical summary of the slopes of different linear regressions between the clade age and species richness and phylogenetic diversity. Ordinary least-square regression analysis (OLS) does not take phylogeny into consideration, while phylogenetic general regression analysis (PGLS) accounts for the phylogenetic relatedness among clades

Contrast	Methods	Slope (*β*)	s.e.	*T*	*p*-value
Clade age versus species richness	OLS	7.475	0.275	27.22	<0.001
	PGLS	5.129	0.716	7.159	<0.001
Clade age versus phylogenetic diversity	OLS	109.63	3.953	27.74	<0.001
	PGLS	0.069	0.0001	428.99	<0.001

Table 2. A comparison of different time-dependent diversification models for fitting the global mammal phylogeny. The best-fit model was marked in boldface

Model	**CONSTANT**	SPVAR	EXVAR
Likelihood	**6191.037**	6187.86	6190.261
AIC	**-12378.07**	-12369.72	-12374.52
Parameters			
λ_0	**0.066**	0.077	0.066
μ_0	**0**	0.001	0.001
k	**-**	0.001	-
z	**-**	-	1.007

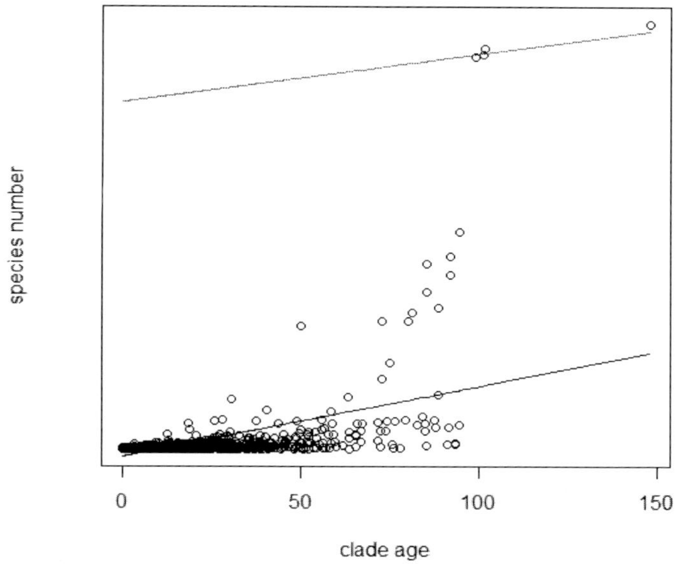

Figure 1. Relationship between clade age and species richness for global mammals. The black fitted line indicated the result from ordinary least-square regression analysis, while the red line is from the phylogenetic general regression analysis.

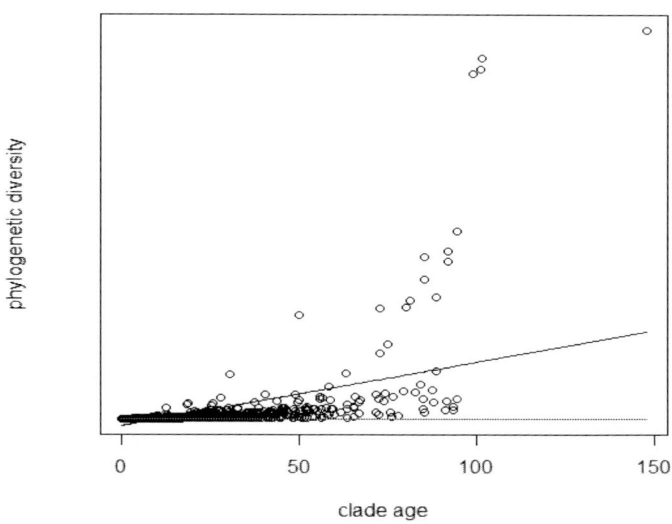

Figure 2. Relationship between clade age and phylogenetic diversity for global mammals. The black fitted line indicated the result from ordinary least-square regression analysis, while the red line is from the phylogenetic general regression analysis.

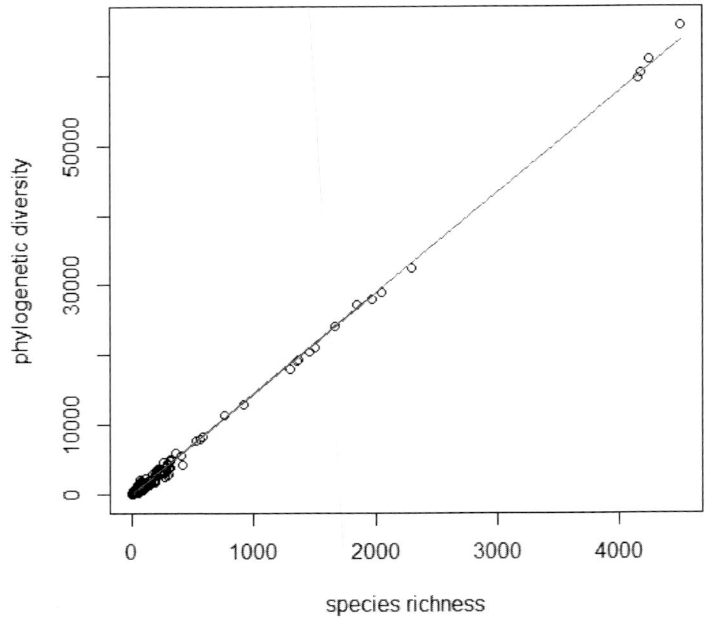

Figure 3. The relationship of species richness and phylogenetic diversity over the internal nodes of global mammal phylogeny.

DISCUSSION

Species diversity is thought to be driven by multiple reasons. One classical hypothesis is that when species have older evolutionary history, they tended to become diversified due to the fact that they have longer time to process speciation events (Etienne et al. 2012; Rabosky et al. 2012). This has been used to explain the mega-diversity of insects in comparison to other large-bodied species (Etienne et al. 2012). Given the current observation for global mammals, the hypothesis was strong supported for two reasons. First, it is observed that there is a significantly positive relationship between clade age and species richness/phylogenetic diversity; second, it is observed that the temporal diversification rates of species are constant over the phylogeny (Table 1), implying that diversification rate-shifting patterns among clades are unlikely. As such, variation caused by varying-diversification rates over time (Rabosky 2009; Rabosky et al. 2012) could not be used to predict species richness and diversity.

In conclusion, the present study showed that global mammalian richness and diversity is explained by clade age, consistent to some previous studies (McPeek and Brown 2007). No significant diversification rate-shifting trends were identified over the evolutionary time.

REFERENCES

Belmaker J, Jetz W (2012) Regional pools and environmental controls of vertebrate assemblages. *Am. Nat.* 179:512–523.

Bininda-Emonds O, Cardillo M, Jones K, MacPhee R, Beck R, Grenyer R, Price S, Vos A, Gittleman J, Purivs A (2007) The delayed rise of present-day mammals. *Nature* 446:507–512.

Cooper N, Freckleton RP, Jetz W (2011) Phylogenetic conservatism of environmental niches in mammals. *Proc. R. Soc. B Biol. Sci.* 278:2384–91.

Davies T, Buckley L (2012) Exploring the phylogenetic history of mammal species richness. *Glob. Ecol. Biogeogr.* 21:1096–1105.

Etienne R, Visser S de, Janzen T, Olsen J, Olff H, Rosindell J (2012) Can clade age alone explain the relationship between body size and diversity? *Interface Focus* 2:170–179.

Grenyer R, Orme CDL, Jackson SF, Thomas GH, Davies RG, Davies TJ, Jones KE, Olson VA, Ridgely RS, Rasmussen PC, Ding T-S, Bennett PM, Blackburn TM, Gaston KJ, Gittleman JL, Owens IPF (2006) Global distribution and conservation of rare and threatened vertebrates. *Nature* 444:93–6.

Jetz W, Fine P (2012) Global gradients in vertebrate diversity predicted by historical area-productivity dynamics and contemporary environment. *PLoS Biol.* 10:e1001292.

McPeek M, Brown J (2007) Clade age and not diversification rate explains species richenss among animal taxa. *Am. Nat.* 169:E97–106.

Rabosky D (2006a) Likelihood methods for detecting temporal shifts in diversification rates. *Evolution (*N. Y). 60:1152–1164.

Rabosky D (2006b) LASER: a maximum likelihood toolkit for detecting temporal shifts in diversification rates from molecular phylogenies. *Evol. Bioinform.* Online 2:247–250.

Rabosky DL (2009) Ecological limits and diversification rate: alternative paradigms to explain the variation in species richness among clades and regions. *Ecol. Lett.* 12:735–43.

Rabosky D, Adams D (2012) Rates of morphological evolution are correlated with species richness in salamanders. *Evolution* (N. Y). 66:1807–1818.

Rabosky D, Lovette I (2008a) Density dependent diversification in North American wood-warblers. *Proc. R. Soc. B Biol. Sci.* 275:2363–2371.

Rabosky D, Lovette I (2008b) Explosive evolutionary radiations: decreasing speciation or increasing extinction through time? *Evolution* (N. Y). 62:1866–1875.

Rabosky D, Slater G, Alfaro M (2012) Clade Age and Species Richness Are Decoupled Across the Eukaryotic Tree of Life. *PLoS Biol.* 10:e1001381.

Safi K, Cianciaruso M, Loyola R, Brito D, Armour-Marshall K, Diniz-Filho J (2011) Understanding global patterns of mammalian functional and phylogenetic diversity. *Philos. Trans. R. Soc. B* 366:2536–2544.

Chapter 7

Spatial Point Pattern Analysis of County Locations of Mainland China: The Relative Roles of Anthropogenic Factors, Climatic Fluctuation and Water Availability

Abstract

In the present study, the distributional patterns of politically administrative units in terms of counties of mainland China are analyzed using spatial point pattern statistical methods. As evidenced by *G*-function with 95% confidence envelope, the county distribution in mainland China is spatially aggregated at small and large spatial scales, while is spatially regular at intermediate spatial scale. This result is partially consistent to my initial prediction that spatial aggregation of the counties over spatial scales should be prevailing. Climatic fluctuation was the most important predictor of county distribution, followed by anthropogenic impacts. Water availability has the least influence on county site selection in mainland China. Latitude and longitude are both equivalently important on structuring the geographic patterns of county distribution in mainland China. The strongest influence caused by climatic variables implies that global change would have deep impacts on the county site selection in the future.

Keywords: socioeconomic analysis, geographical information system, urbanization, spatial statistics

INTRODUCTION

As the largest developing country over the world, China is undergoing rapid urbanization and industrialization (Kojima 1995; Zhou et al. 2004). In the next decade, China will relocate its rural population as much as possible into counties and cities (http://en.ndrc.gov.cn/). By the end of 2025, it is estimated that over 200 cities of China will have over one million residents (McKinsey Global Institute 2009). The high speed of urbanization might pose great impacts onto environmental protection (Zhu 2012), people's health (Zhu et al. 2011) and climate (Zhou et al. 2004) issues for China.

City and county distribution for a nation is driven by multifaceted factors. Most likely, the allocation and setup of a county in a region is determined by governmental and political decision, which will consider human population size and industrialization degree of the county. However, city and county distribution are not fully controlled by human society because water availability and climatic condition should be also dominant. It is easily expected that no or just a few counties or cities would be set up in those harsh (extremely cold or dry) areas, regardless of the nations around the world. Thus, it might be of some interests to study the contemporary distribution of counties/cities and reveal the underlying environmental and socioeconomic processes on structuring the distribution of these politically administrative units.

I have the following hypotheses for linking the spatial point patterns of county distribution in mainland China with climate, landscape and agriculture. First, although both latitude and longitude reflect the importance of climatic variability, habitat heterogeneity and water availability for human residence, longitude is predicted to be a more important variable. A major reason is the availability of water resources. In western part of China, there is a huge area of desert which is lacked of water. As such, people tend to avoid living there. Moreover, the whole territory range of mainland China is still within the temperate zone, thus, even though the temperature and precipitation are greatly different between northern and southern part of China. The influence of such kinds of differences is not as high as that for water availability (rivers, lakes and others) on the historical residence of Chinese people.

Second, as seen in Figure 1, most of the counties are located in eastern part of China, it apparently presents aggregation patterns. Such a formation of aggregation distribution of counties in China can be driven by various factors, for example, the river resources, the transportation convenience and the agricultural farmlands. As such, I predict that the county distribution in mainland region of China should follow the aggregation in comparison to complete random spatial distribution (CSR) or regular patterns. I will use a series of spatial point statistical methods to test this prediction, including *G*-function.

At last, because human activities are very crucial in the site selection of counties at both ancient and contemporary histories, I predict that the anthropogenic factors are one of the most important predictors of county distribution in mainland China. For quantifying the impacts of anthropogenic factors, I used both cropland distribution (which indicates the agricultural farming) and human footprint index (which synergistically incorporates the information of population size, infrastructure and others) as the indicators.

Spatial point analysis is a commonly used tool in ecological and epidemiological studies for revealing the distributional patterns of species (Diggle 2003; Baddeley and Turner 2005; Eckel 2008; Gray and He 2009). However, it has not been widely used along with geographic information system yet and is restrictedly used in plant ecology (Shen et al. 2009, 2013). It has the capability to reveal the underpinning mechanisms structuring geographic patterns of spatial points (Diggle 2003; Baddeley and Turner 2005; Shen et al. 2013).

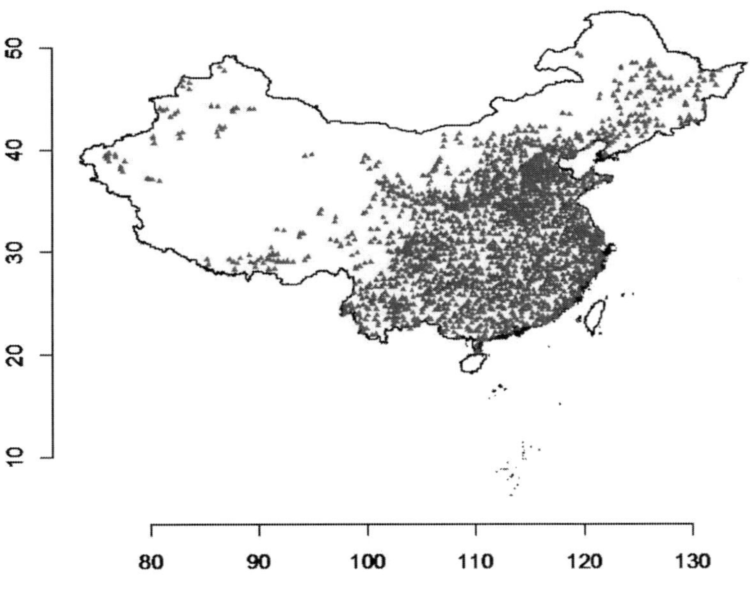

Figure 1. Geographic locations of 2046 counties in mainland China.

In the present study, by incorporating both spatial point pattern analysis and geographic information system, I am able to get a better prediction on contemporary county distribution in mainland terrestrial region of China and the possible elucidation of the relative importance of climate, landscape heterogeneity, water availability and human footprints on determining the geographic patterns of counties.

MATERIALS AND METHODS

County Distribution in Mainland Region of China

The geographic coordinates of 2049 counties in China are collected from National Geomatics Center of China (http://ngcc.sbsm.gov.cn/). Each county contains the information of the name and geographic coordinates of the location.

As seen, after excluding counties presented in Taiwan and Hainan Islands, there are 2048 counties remained in mainland China (Figure 1). As such, the present study is only interested into the spatial point pattern analysis of county distribution in mainland territory of China.

Socioeconomic and Environmental Predictors

I collect the following socioeconomic variables for studying their correlations with the county locations in mainland China, including human footprint index and cropland density. I use annual minimum, maximum and mean temperatures as the indicators of climatic fluctuation and annual precipitation as the indicator of water availability.

G Function

G function is a classical spatial statistic for measuring the point randomness patterns, which is defined as the probability that the distance from any random event to its nearest neighbor is less than or equal to a given distance s:

$$G(s) = \sum_{i=1}^{n} I(d_i \le s) / n \tag{1}$$

Where n is the number of total observed distributional points, $I(\bullet)$ is the indicator function; d_i is the distance between the i-the spatial point and its nearest neighbor. When the distance d_i is smaller than or equal to the given distance s, $I(d_i \le s)$ =1. Under complete spatial randomness (CSR), the expected K function is $G(s)_{CSR} = 1 - e^{-\lambda \pi s^2}$ (2). Here, λ is the density or intensity of the observed points over the sampling area.

Spatial Point Process Models

Both homogeneous and inhomogeneous models are analyzed for the present study, including homogeneous and inhomogeneous Poisson process models.

Homogeneous Poisson Process (HPP)

HPP model assumes a constant intensity value λ for the spatial points over the whole concerned area. That is, the intensity of spatial point events is independent of spatial locations (Model 0). HPP model is served as the null model and compared to other IPP models as described as follows.

Inhomogeneous Poisson Process (IPP)

IPP model assumes that the intensity is dependent on the spatial locations. That is, $\lambda(u)$ is a function of geographic locations u.

The following simple IPP models are implemented:

(1) λ is a function of longitude (x) only: $\lambda(u) = \exp(\theta_0 + \theta_1 x(u))$ (Model 1).

(2) λ is a function of latitude (y) only: $\lambda(u) = \exp(\theta_0 + \theta_1 y(u))$ (Model 2).

HPP is served as the null model. If HPP is not rejected, the distribution of spatial points for the species is thus should be inhomogeneous. Model selection can be fulfilled by using Akaike Information Criteria (AIC) values across the models. Lower AIC a model has, better the model is.

Quantifying the Relative Influence of Human Footprints, Climatic Variability and Water Availability

I analyzed and compared alternative IPP models to further evaluate the relative importance of human footprints, climatic variability and water availability on structuring the spatial geographic patterns of county locations in mainland China.

(3) λ is a function of annual mean, minimal, maximal temperature (T1, T2 and T3) and annual mean precipitation (P): $\lambda(u) = \exp(\theta_0 + \theta_1 T_1(u) + \theta_2 T_2(u) + \theta_3 T_3(u) + \theta_4 P(u))$ (Model 3: quantifying the influence of climatic variability).

(4) λ is a function of relative wetness (W) and the nearest distance to the river (R): $\lambda(u) = \exp(\theta_0 + \theta_1 W(u) + \theta_2 R(u))$ (Model 4: quantifying the influence of water availability).

(5) λ is a function of human footprint index (H) and cropland density (C): $\lambda(u) = \exp(\theta_0 + \theta_1 H(u) + \theta_2 C(u))$ (Model 5: quantifying the influence of anthropogenic factors).

Again, I utilize AIC values to choose the best model. There is significant difference between two models only when the absolute difference of AIC values between two models is larger than 10. All the above-mentioned statistical analyses are done using "*spatstat*" package (Baddeley and Turner 2005) and some self-made scripts under R statistical computing environment (R Development Core Team 2013).

RESULTS

As showed in Figure 2, the G-function graph indicated that the county distribution in mainland China was aggregated at small and large distances (*r*<0.3 and *r*>0.6) because at both distance intervals the observed curve (black line) was fallen below the lower bound (grey dashed line) of the 95% confidence zone for the theoretical Poisson CSR curve (red line).

However, at intermediate distances (0.3<*r*<0.6), the observed G curve was above the higher bound of the 95% confidence zone for the CSR curve, implying that at this spatial distance interval, county distribution presented spatially regular patterns.

Through model selection, as showed in Table 1, both latitude and longitude are equally important on determining the spatial distribution patterns of counties in mainland China. Both models (Model 1 and 2) have significant lower deviance values in comparison to that in CSR model (Model 0). Moreover, when comparing the deviance values between Model 1 and 2, no significance was detected (deviance=-8.12), implying that both models are not different from each other. As such, this result implied both latitude and longitude are equivalently important predictors of geographic locations of counties in mainland China.

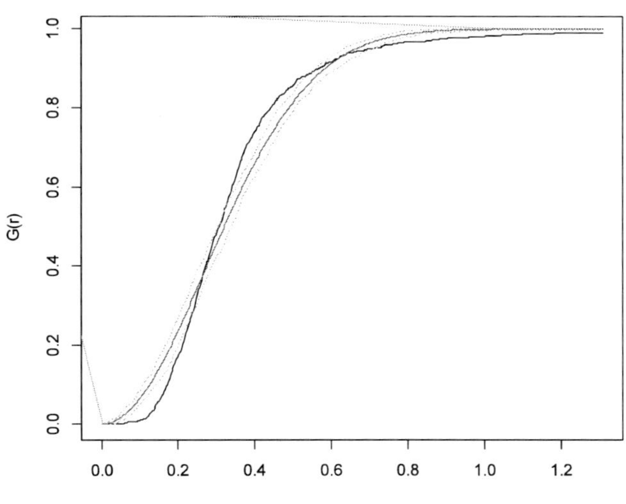

Figure 2. *G* function curve for the county distribution in mainland China. Black line indicates the observed *G* function curve, while red line indicates the theoretical Poisson CSR curve. 95% confidence curves are denoted by dashed gray lines.

Table 1. Model comparison between IPP and HPP models for quantifying the relative influence of latitude and longitude on the distribution of counties in China

Model comparison	Δ AIC	*P* value
Model 0 - Model 1	-536.99	<0.05
Model 0 - Model 2	-528.9	<0.05
Model 1 - Model 2	-8.12	>0.05

Table 2. Model comparisons between IPP models accounting for the impacts of climatic variability, anthropogenic factors and water availability on structuring distributional patterns of counties in mainland China. Variables with significant coefficients are marked by asterisks

Model	Estimated parameters	AIC value
Model 3	$\lambda(u) = \exp(2.71 + 0.03T_1(u) - 0.007T_2(u) - 0.01T_3(u) - 0.01P(u)^*)$	-401.53
Model 4	$\lambda(u) = \exp(0.56 - 2.33W(u)^* + 0.01R(u)^*)$	-24.08
Model 5	$\lambda(u) = \exp(-54 + 0.70H(u)^* + 0.04C(u)^*)$	-324.69

Through model selection and comparison, as showed in Table 2, climatic variability is much more preferred in comparison to other groups of variables on structuring the distribution of counties in mainland China, which has the lowest AIC value=-401.53. In contrast, In contrast, model 4 with relative wetness and distance to the rivers received a very high AIC value, indicating that it was very unfavoured.

DISCUSSION

Compared to those modern cities and/or counties in North America, which have been set up no more than 400 years ago (Barra Foundation 1982; Avery 1999; Homberger and Hudson 2005), the establishment of each county in mainland China is full of historical information and cultural background. The present study utilized spatial point pattern analysis to analyze the county distribution pattern of mainland China and suggested that both latitude and longitude are equivalently important on determining the county locations.

Surprisingly, both latitude and longitude are equally important predictors for the county location in China. The reasons might be due to the fact that the setup of county is principally determined by political decisions and population size of the county. Thus, it is a completely socioeconomic activity rather than a biological diffusion process.

I choose cropland density as an additional index for quantifying anthropogenic factors because it indicates human agricultural activities (Ramankutty and Foley 1999; Li et al. 2010). The historical cropland patterns are in concordance with the history of human settlement and patterns of economic development over the areas (Ramankutty and Foley 1999; Li et al. 2010). Thus, usage of cropland information might be a way to reflect historical human activities, which thus can be used to model how human footprint would structure the current county locations around China. However, based on AIC values, it was found that cropland density itself was a poor indicator of county location in mainland China. Such a result might be due to that fact that most available croplands are usually located in rural areas, which have some distances from the county centers. More importantly, the alternation of croplands over the historical time is remarkable as indicated by some previous studies (Ramankutty and Foley 1999; Li et al. 2010). Therefore, utilization of contemporary cropland map cannot recover the historical cropland change and associated footprints of human socioeconomic activities (Goldewijk 2001; Lepers et al. 2005). One needs to reconstruct historical cropland patterns (Ramankutty and Foley 1999; Goldewijk 2001) so as to better model the dynamic of county distribution.

Being discordant with my initial predictions and perceptions, anthropogenic factors were not the best predicator of county location, and spatial distribution patterns of counties in mainland China were not aggregate but regular at both small and large spatial scales. Climatic variability is more important to influence the geographic locations of counties might be a signal of historical influence. In ancient time, human beings had to migrate across various areas to find suitable residence places with benign climate. Moreover, because of competition among tribes, they tend to avoid each other and explore new habitats with warm climatic condition. These activities make the climatic variability is closely related to the residence of human beings in ancient China.

The regular patterns of site selection of counties might be related to that political consideration: across different dynasties, the setup of a political administrative unit in China typically would consider its transportation efficiency, supply convenience, military control to/from lower-level administrative units (like villages and small towns) and higher-level ones (like city and capital) (Hou 1979; Li 2010; Cheng 2012). Such kinds of decision-making processes may drive policy makers to choose a site around the center of a focused area as the location of a county, resulting into the regular pattern of county distribution in mainland China.

Interestingly, water availability plays the least influence on structuring the locations of counties based on the present study. At both historical and current time, water resource is always one of the most important factors to set up a county or town (Hou 1979; Cheng 2012). It might be due to that fact that my analysis covers the broad spatial extent, obscuring the importance of water resource availability on structuring county spatial distribution patterns in mainland China. However, without doubt, the influence of water resources on county site planning is of top priorities at local spatial scale.

My study implied that there was a close association between climatic condition and county distribution. Different from a previous study (Zhou et al. 2004), which predicted that urbanization would increase the ground temperature in China, my present study showed that climatic condition can predict the site selection of counties in China. As global climate warming is accelerated, when the sea level rises to a higher level, the relocation of those counties along the coastal areas of eastern part of China to higher-elevation areas might be expected (Han et al. 1995). In such a context, the relation between climate and county geographic locations would be strengthened.

REFERENCES

Avery R (1999) A concise history of Philadelphia. *Otis Books.*

Baddeley A, Turner R (2005) Spatstat: an R package for analyzing spatial point patterns. *J. Stat. Softw.* 12:1–42.

Barra Foundation (1982) Philadelphia: a 300-year history. Barra Foundation.

Cheng Y (2012) Research methods of ancient Chinese cities site selection. *Collect. Essays Chinese Hist. Geogr.* 27:84–93.

Diggle P (2003) Statistical analysis of spatial point patterns. Academic Press, London, UK.

Eckel SM (2008) Statistical Analysis of Spatial Point Patterns: Applications to Economical, Biomedical and Ecological Data.

Goldewijk K (2001) Estimating global land use change over the past 300 years: the HYDE database. Global Biogeochem. *Cycles* 15:417–433.

Gray L, He F (2009) Spatial point-pattern analyssi for detecting density-dependent competition in a boreal chronosequence of Alberta. *For. Ecol. Manage.* 259:98–106.

Han M, Hou J, Wu L (1995) Potential impacts of sea-level rise on China's coastal environment and cities: a national assessment. *J. Coast. Resarch* 14:79–95.

Homberger E, Hudson A (2005) The historical atlas of New York City: a visual celebration of 400 years of new york city's history. *Owl Books.*

Hou R (1979) The study of urban historical geography and city planning. *Acta Geogr. Sin.* 34:315–328.

Kojima R (1995) Urbanization in China. *Dev. Econ.* 33:151–154.

Lepers E, Lambin E, Janetos A, DeFries R, Achard F, Ramankutty N, Scholes R (2005) A synthesis of information on rapid land-cover change for the period 1981-2000. *Bioscience* 55:115–124.

Li M (2010) Essential factor analysis to the site selection for the administration center of the minority area. *Asian Soc. Sci.* 6:24–27.

Li B, Fang X, Ye Y, Zhang X (2010) Accuracy assessment of global historical cropland datasets based on regional reconstructed historical data-A case study in Northeast China. *Sci. China Earth Sci.* 53:1689–1699.

McKinsey Global Institute (2009) Preparing for China's urban billion.

R Development Core Team (2013) R: A Language and Environment for Statistical Computing, Vienna, Austria. ISBN 3-900051-07-0, URL http://www.R-project.org.

Ramankutty N, Foley J (1999) Estimating historical changes in global land cover: croplands from 1700 to 1992. *Global Biogeochem. Cycles* 13:997–1027.

Shen G, He F, Waagepetersen R, Sun I, Hao Z, Chen Z, Yu M (2013) Quantifying effects of habitat heterogeneity and other clustering processes on spatial distributions of tree species. *Ecology:In press.*

Shen G, Yu M, Hu X, Mi X, Ren H, Sun I, Ma K (2009) Species-area relationships explained by the joint effects of dispersal limitation and habitat heterogeneity. *Ecology* 90:3033–3041.

Zhou L, Dickinson R, Tian Y, Fang J, Li Q, Kaufmann R, Tucker C, Myneni R (2004) Evidence for a significant urbanization effect on climate in China. *PNAS* 101:9540–9544.

Zhu Y (2012) Environmental impacts of rapid urbanization in China: a showcase of recent research developments. *Environ. Sci. Pollut. Res.* 19:1351.

Zhu Y, Ioannidis J, Li H, Jones K, Martin F (2011) Understanding and hamessing the health effects of rapid urbanization in China. *Environ. Sci. Technol.* 45:5099–5104.

Chapter 8

MODELING SPECIES-AREA RELATIONSHIP WITH MEASUREMENT UNCERTAINTY

ABSTRACT

In the present study, the role of measurement uncertainty of species richness has been taken into account when estimating the parameters (c and z) for the power-law species-area relationship (SAR). The nonlinear weighted estimator χ^2 is used to quantify the influence of measurement uncertainty in the values of species richness, which can be derived from the observed and estimated species richness across different areas. As a comparison, the parameters are also estimated using the conventional nonlinear least-of-square (NLOS) estimator without considering data uncertainty and only the average species richness from estimated and observed values is used. Species richness for epigean arthropods (EAR), canopy arthropods (CAJ) and ground bryophytes (BD) over different areas at the Azores, Portugal are used as empirical data sets for comparing the proposed χ^2 and conventional NLOS estimators. The results show that, both parameters c and z estimated by χ^2 estimator are significantly different from those from NLOS respectively through the paired t-test in all the three empirical data sets except that c values are not significantly different for the BD data set when comparing both estimators. Given that fact that there are significant differences on the estimated parameters for the power-law SAR model when comparing both estimators, χ^2 estimator is recommended for fitting SAR models so as to better capture the stochasticity of species richness.

Keywords: ecological scaling, small island effect, extinction risk, measurement errors, statistical inference

INTRODUCTION

Species-area relationship (SAR) is one of the most classical ecological laws (Arrhenius 1921; He and Legendre 1996). SAR has been widely used to estimate and predict species' extinction risk (Halley et al. 2013; He and Hubbell 2013; He and Hubbell 2012; He and Hubbell 2011; Kinzig and Harte 2000). A variety of SARs has been invented and applied to cope with the influences of habitat diversity, landscape heterogeneity and island age (Cameron et al. 2013; Chen 2009; Hortal et al. 2009; Patino et al. 2013; Sfenthourakis and Triantis 2009; Triantis et al. 2008).

Limited sampling efforts have been broadly observed in ecological studies (Chase and Knight 2013; Hortal 2012). As such, using the inventory data of species richness across different areas as the representative to estimate the slopes of SARs might tend to over- or under- represent the true SAR patterns. This discrepancy based on limited sampling efforts can be further exacerbated when landscape and/or dispersal complexity are taken into consideration as previously described (Hu et al. 2012; Kolasa et al. 2012; Plotkin et al. 2000; Shen et al. 2009).

Small island effect is another factor that might bring more uncertainty into the estimation of SAR parameters (Hortal 2012; Sfenthourakis and Triantis 2009; Wang et al. 2012). SIE is a hypothesis stating that the species richness of islands would become independent on the areal sizes of the corresponding islands (Wang et al. 2012). As such, when plotting species richness over area sizes, data points at the zone where small areas are located would be much over dispersed based on the prediction of SIE.

As such, it seems very necessary to develop new statistical estimators to take into account of the measurement uncertainty of species richness observation over different areas so as to accurately evaluate the slopes of SAR curves. In the present study, the simple χ^2 statistic is used to control the measurement uncertainty in the species richness of a given island. The measurement uncertainty here denotes the difference of observed and estimated species richness for the focused area, which is calculated from different richness estimators.

MATERIALS AND METHODS

Statistical Methods

The nonlinear weighted estimator χ^2 has been widely applied in different disciplines including biology and physics (Ma 2008). Supposing that there is a data point *(x,y)* for constructing a SAR model $S = f(A)$, we assume here that *x* denotes the size of areas, while *y* denotes the species number of that area. To incorporate the measurement uncertainty, we quantify the uncertainty of species richness as the standard deviation (δ_y) of the species richness based on the observed and estimated values using alternative richness estimators (Colwell 2006; Hortal et al. 2006). Then, the χ^2 statistic is the summation of the quantity $\frac{(y-f(x))^2}{\delta_y^{\ 2}}$ over the sampling areas as follows,

$$\chi^2 = \sum_i \frac{(y_i - f(x_i))^2}{\delta_{y_i}^{\ 2}} \tag{1}$$

Where *i* is the index of a sampling area. Minimizing the above index allows ones to obtain the estimation of parameters related to the SAR model $S = f(A)$ while considering the influence of measurement uncertainty of species richness.

When $f'(x)$ is not a constant (i.e., not a linear model), equation (1) can be regarded as the nonlinear weighted least-squares estimates of the parameters of a nonlinear model. Thus, the numerical optimization technique should be applied to estimate the relevant parameters.

The conventional nonlinear least-of-square estimator (NLOS) is implemented as a comparison. The NLOS is simply to ignore the standard deviations of the data points as follows,

$$NLOS = \sum_i (y_i - f(x_i))^2 \tag{2}$$

Data Sets and the SAR Model

Three empirical datasets are obtained from a previous study (Borges et al. 2009), which include a data set for the soil epigean arthropods at eight forest fragments in Terceira Island (named as EAR dataset), a data set for the canopy arthropods inhabiting *Juniperus brevifolia* at sixteen forest fragments of six different islands (named as CAJ dataset), and a data set for the bryophytes of seven forest fragments from Terceira and Pico islands (named as BD dataset). These data sets are built form the inventory of species diversity of arthropods and bryophytes at the Azores, Portugal (Borges et al. 2009).

In that previous study (Borges et al. 2009), the observed species richness has been provided and the associated estimated richness for each area of the data sets has been calculated using a variety of non-parametric richness estimators (ACE, ICE, Chao1, Chao2, Jackknife1, Jackknife2 and Bootstrap). The detailed introduction of these methods are not presented here for simplicity and should refer to previous studies if interested (Borges et al. 2009; Colwell 2006; Colwell and Coddington 1994; Fattorini 2009; Gotelli and Colwell 2001; Hortal et al. 2006).

For the present study, the power-law SAR model is utilized for comparative studies. Here, the power-law equation is written as,

$$S = cA^z \tag{3}$$

Where S is the species richness, A the area size, c and z free parameters required to be estimated.

The parameter estimation for c and z is carried out on the original equation (3) without any log-transformation. Transformation of the data may lead to unexpected results (Packard 2013; Packard 2009; Xiao et al. 2011).

RESULTS

As showed in Table 1 and Figures 1A and 1B, for both EAR and CAJ data sets, the estimated parameters c and z values for power-law SAR models are significantly different between the NLOS and χ^2 estimators. The goodness of fit for χ^2 is slightly smaller than that for NLOS estimator because it needs to take into account of the influence of data uncertainty.

For BD data set (Figure 1C), the results are basically similar to those for EAR and CAJ datasets. However, the estimated parameter c values from NLOS and χ^2 are not significantly different from each other.

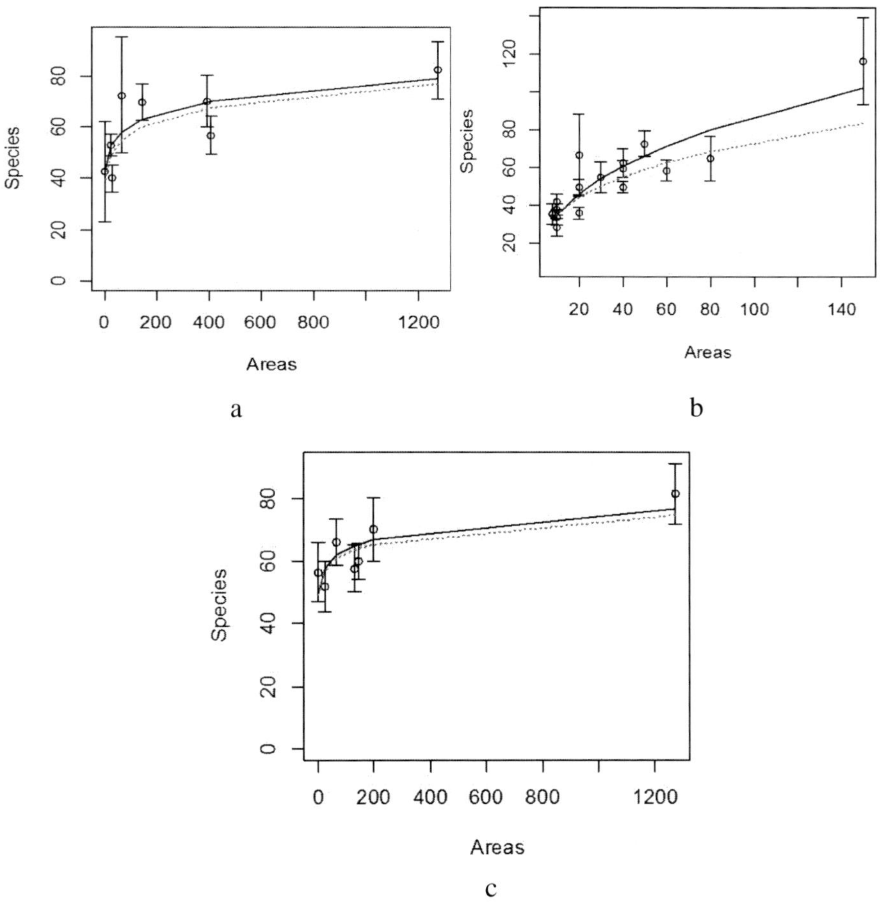

Figure 1. Power-law SAR curves for the species diversity of epigean arthropods (EAR, A), canopy arthropods (B) and ground bryophytes (C) data sets at the Azores, Portugal. Bars on the hollow points indicated the standard deviation of the data. The dashed red line indicated the fitting derived from the proposed χ^2 estimator using standard deviation information of species richness from estimated and observed values, while the solid back line indicated the fitting from traditional nonlinear fitting without taking into account of standard deviation of richness but only the average values of richness.

Table 1. Estimation and comparison of parameters for the power-law SAR models using the conventional nonlinear fitting (NLOS) and proposed χ^2 estimators for the species diversity of epigean arthropods (EAR), canopy arthropods (CAJ) and ground bryophytes (BD) data sets at the Azores, Portugal. *c* and *z* are the parameters in the power-law SAR model. *SE* denotes the standard errors for the estimated parameters. R^2 reflects the goodness of fit of the model. *t* denotes the result from the two-tailed *t*-test, which compares the differences on simulated 1000-pair random values. These random pairs are generated from normal models with the mean = estimated value (*c* or *z*) and the standard deviation = $SE \times \sqrt{n}$ for both estimators. * denotes a significant difference with *p*<0.05

Datasets	Estimators	*c*	*SE*	*t*	*z*	*SE*	*t*	R^2
EAR	NLOS	37.535	7.08		0.104	0.035		0.98
	χ^2	33.772	6.353	4.398*	0.115	0.039	-2.606*	0.98
CAJ	NLOS	13.846	2.852		0.400	0.052		0.974
	χ^2	16.952	2.904	-5.279*	0.318	0.052	8.975*	0.960
BD	NLOS	45.832	5.515		0.071	0.024		0.992
	χ^2	45.389	6.289	0.222	0.069	0.028	2.684*	0.992

DISCUSSION

The importance of data uncertainty has been growingly appraised in recent years for ecological and environmental modeling (Cressie et al. 2009; Falk 2010; Hui and Jackson 2007; Jager and King 2004; Moore and Semmens 2008; Phillips and Gregg 2001). In the present study, the uncertainty of species richness for each area when constructing SAR models has been resolved by using the χ^2 nonlinear weighted least-squares estimator. Based on the comparison between the models with and without measurement uncertainty, it is found that the situations when uncertainty is taken into account would tend to have smaller parameter values (Table 1).

The influence of measurement uncertainty of species richness using the χ^2 estimator is dependent on the standard deviation of species richness data from observed and estimated values for each focused area. As presented in the equation (1), when the standard deviation of species richness is high, the contribution of the data point from the focused area for the overall χ^2 value will become trivial. In contrast, when the standard deviation is low, the contribution of a specific data point for the overall χ^2 value will be remarkable. Thus, the resulting estimated parameters through χ^2 and NLOS estimators are expected to be different since NLOS didn't include uncertainty information. Based on the comparative results on applying both estimators to three empirical data sets in the present study, the significant differences on the estimated parameter values indeed were observed (Table 1).

The purpose of employing nonlinear weighted least-squares estimator χ^2 by incorporating measurement errors is not because of its novelty but simplicity and straightforward understanding. It can be regarded as the extension of the ordinary least-of-square (LOS) minimization technique when considering the influence of data uncertainty. Thus, the χ^2 estimator is nothing new but a weighted version of NLOS estimator. The estimator can be feasibly implemented in numerical optimization and programming coding. Or alternatively, it can be easily implemented under the R computing environment (R Development Core Team 2013) using the function "*nls*" with weights.

At last, the present results should not be directly compared to those presented in the previous study (Borges et al. 2009) for a variety of reasons. First, we take into account of all estimated and observed species richness together when reconstructing the SAR model. In contrast, the previous study estimated SAR models for each of the estimated and observed richness. Second, the present study takes into account the data uncertainty while the previous study (Borges et al. 2009) didn't do that. At last and more importantly, our estimation of SAR model parameters is carried out on the original power-law equation (3) without log-transformation. However, the previous study (Borges et al. 2009) estimated all free parameters using the log-transformed SAR model. Thus, since that there is a continuous debate on whether log-transformation of the data should be applied (Packard 2013; Packard 2009; Xiao et al. 2011), it is unwise to directly compare the results presented here and the previous work.

REFERENCES

Arrhenius O (1921) Species and area. *J Ecol* 9:95–99.

Borges P, Hortal J, Gabriel R, Homem N (2009) Would species richness estimators change the observed species area relationship? *Acta Oecologica* 35:149–156.

Cameron R, Triantis K, Parent C, et al. (2013) Snails on oceanic islands: testing the general dynamic model of oceanic island biogeography using linear mixed effect models. *J Biogeogr* 40:117–130.

Chase J, Knight T (2013) Scale-dependent effect sizes of ecological drivers on biodiversity: why standardised sampling is not enough. *Ecol Lett 16*:17–26.

Chen Y (2009) Combining the species-area-habitat relationship and environmental cluster analysis to set conservation priorities: a study in the Zhoushan Archipelago, China. *Conserv Biol* 23:537–545.

Colwell R (2006) EstimateS: statistical estimation of species richness and shared species from samples. Version 8.

Colwell R, Coddington J (1994) Estimating terrestrial biodiversity through extrapolation. *Philos Trans R Soc Lond B Biol Sci* 345:101–118.

Cressie N, Calder C, Clark J, et al. (2009) Accounting for uncertainty in ecological analysis: the strengths and limitations of hierarchical statistical modeling. *Ecol Appl* 19:553–570.

Falk M (2010) Incorporating uncertainty in environmental models informed by imagery. PhD thesis, Queensland University of Technology.

Fattorini S (2009) Darkling beetle communities in two geologically contrasting biotopes: testing biodiversity patterns by microsite comparisons. *Biol J Linn Soc* 98:787–793.

Gotelli N, Colwell R (2001) Quantifying biodiversity: procedures and pitfalls in the measurement and comparison of species richness. *Ecol Lett* 4:379–391.

Halley J, Sgardeli V, Monokrousos N (2013) Species-area relationships and extinction forecasts. A*nn N Y Acad Sci* 1286:50–61.

He F, Hubbell S (2011) Species–area relationships always overestimate extinction rates from habitat loss. *Nature* 473:368–371.

He F, Hubbell S (2012) Geometry and scale in species-area relationshps; Extinction and climate change: a reply. *Nature* 482:E5–E6.

He F, Hubbell S (2013) Estimating extinction from species-area relationships: why the numbers do not add up? *Ecology* In press.

He F, Legendre P (1996) On species-area relations. *Am Nat* 148:719–737.

Hortal J (2012) Escaping the trap of low sample size in island biogeography. *Front Biogeogr* 3:4.

Hortal J, Borges P, Gaspar C (2006) Evaluating the performance of species richness estimators: sensitivity to sample grain size. *J Anim Ecol* 75:274–287.

Hortal J, Triantis K, Meiri S, et al. (2009) Island species richness increases with habitat diversity. *Am Nat* 174:E205–E217.

Hu G, Wu J, Feeley K, et al. (2012) The effects of landscape variables on the species-area relationship during late-stage habitat fragmentation. *PLoS One* 7:e43894.

Hui D, Jackson R (2007) Uncertainty in allometric exponent estimation: a case study in scaling metabolic rate with body mass. *J Theor Biol* 249:168–177.

Jager H, King A (2004) Spatial uncertainty and ecological models. Ecosystems 7:841–847.

Kinzig A, Harte J (2000) Implications of endemics-area relationships for estimates of species extinctions. *Ecology* 81:3305–3311.

Kolasa J, Manne L, Pandit S (2012) Species-area relationships arise from interaction of habitat heterogeneity and species pool. *Hydrobiologia* 685:135–144.

Ma Y (2008) Variable cosmological constant model: the reconstruction equations and constraints from current observational data. *Nucl Phys B* 804:262–285.

Moore J, Semmens B (2008) Incorporating uncertainty and prior information into stable isotope mixing models. *Ecol Lett* 11:470–480.

Packard G (2009) On the use of logarithmic transformations in allometric analyses. *J Theor Biol* 257:515–518.

Packard G (2013) Is logarithmic transformation necessary in allometry? *Biol J Linn Soc DOI:* 10.1111/bij.12038.

Patino J, Guilhaumon F, Whittaker R, et al. (2013) Accounting for data heterogeneity in patterns of biodiversity: an application of linear mixed effect model to the oceanic island biogeography of spore-producing plants. *Ecography* (Cop) In press.

Phillips D, Gregg J (2001) Uncertainty in source partitioning using stable isotopes. *Oecologia* 127:171–179.

Plotkin JB, Potts MD, Leslie N, et al. (2000) Species-area curves, spatial aggregation, and habitat specialization in tropical forests. *J Theor Biol* 207:81–99. doi:10.1006/ jtbi.2000.2158

R Development Core Team (2013) R: A Language and Environment for Statistical Computing, Vienna, Austria. ISBN 3-900051-07-0, URL http://www.R-project.org.

Sfenthourakis S, Triantis K (2009) Habitat diversity, ecological requirements of species, and the small island effect. *Divers Distrib* 15:131–140.

Shen G, Yu M, Hu X, et al. (2009) Species-area relationships explained by the joint effects of dispersal limitation and habitat heterogeneity. *Ecology* 90:3033–3041.

Triantis K, Mylonas M, Whittaker R (2008) Evolutionary species-area curves as revealed by single-island endemics: insights for the interprovincial species-area relationship. *Ecography* (Cop) 31:401–407.

Wang Y, Zhang M, Wang S, et al. (2012) No evidence for the small island effect in avian communities on islands of an inundated lake. *Oikos* 121:1945–1952.

Xiao X, White E, Hooten M, Durham S (2011) On the use of log-transformation vs. nonlinear regression for analyzing biological power laws. *Ecology* 92:1887–1894.

Chapter 9

ESTIMATING POPULATION CHANGE OF SPECIES UNDER CLIMATE CHANGE USING OCCUPANCY-ABUNDANCE RELATIONSHIP

ABSTRACT

Quantitatively evaluating biodiversity change required us to know the shifting pattern of species population and the distributional range. In the current study, by using occupancy-abundance relationship and considering the aggregation of species distributional patterns, I examined the change of species distributional ranges and abundance under future climatic change in comparison to the situations at current climatic scenarios. 93 endemic amphibians in China were considered as the case study. The results showed occupancy-abundance relationship could be either positive or negative, depending on the estimation methods of abundance we used. The utilization of occupancy-abundance relationship would provide a way to estimate species abundance. Partial canonical correspondence analysis suggested that climate was the dominant factor influencing species' range occupancy (explaining over 50% of total variance), and the interaction between climate and space further explained over 15% of total variance in species occurrence. However, pure space affected species' presence trivially. The negative occupancy-estimated abundance relationship using the fully random non-aggregation model might be derived from the decoupling of species' occurrence probability predicted by ecological niche models and the real occupied grid cell numbers of species across the landscape.

Keywords: species distribution, biodiversity crisis, global change, species-environment relationships

INTRODUCTION

Abundance is regarded as a vital ecological feature of a species (Figueiredo and Grelle 2009), which has broad implications in species conservation and assessment of species threatened status (He 2012). However, in many real applications, the abundance of threatened species is hard to know due to limited sampling effects and financial budgets (He 2012). As such, estimation of species threatened status should rely on approximate approaches. In some recent works, one important progress (He 2012) of these alternative approaches have been initialized by using occupancy information of species (He and Gaston 2000; He 2012).

Ecological niche modeling (ENM) is a widely applied technique to investigate fundamental and realized niches of species, which were subjected to complex ecological and evolutionary dynamics (Liu et al. 2012). In recent years, ENM have been applied to model and predict species distribution for conservation monitoring (Urbina-Cardona and Flores-Villela 2010), reveal invasive success of alien species (Peterson and Vieglais 2001; Rodder et al. 2009; Venette et al. 2010), predict the outbreak of epidemic diseases (Peterson et al. 2002; López-Cárdenas et al. 2005), understand the niche differentiation and shifting of species under climate change (Medley 2010; Soberon and Peterson 2011; Petitpierre et al. 2012) and help delimitate new species (Raxworthy et al. 2007; Hawlitschk et al. 2011; Reeves and Richards 2011).

Traditionally, a positive occupancy-abundance relationship was observed and predicted by both empirical and theoretical studies (He and Gaston 2000, 2007; Webb et al. 2012). However, negative possibilities could not be omitted as well (Gaston 1996; Gaston et al. 2000; Blackburn et al. 2006; Frisk et al. 2011; Wilson 2011). The reasons of negative relationships were unclear, but should be interesting since there may be some hidden biological mechanisms involved (Webb et al. 2012).

Habitat heterogeneity should be one of factors that may contribute to species occupancy (Hurlbert and White 2007). For both generalists and specialists, they would behave differently for high-heterogeneous landscapes. Because generalists could occur anywhere, while specialists should only colonize part or a few suitable areas due to resource distributional aggregation (Hurlbert and White 2007), such a habitat preference differentiation may result into a closed association between habitat heterogeneity and species' occupancy.

Another factor that may contribute to species' range occupancy but has not been mentioned in previous studies is spatial autocorrelation. Spatial autocorrelation is ubiquitous in spatial analysis (Legendre 1993; Dormann 2007), which showed that distance-closed sites should have high level of similarity or occurrence probability of species, while distance-distinct sites low level of similarity or occurrence probability. As such, we would expect that space should play a role in structuring species' occupancy.

In the current study, I utilized occupancy-abundance relationship (He et al. 2002; Hui et al. 2012) to estimate species abundance under current and future climate scenarios and predict the change of species abundance. The purpose of the study is to offer an evaluation method of biodiversity change happened at population level. The method could help conservationists to better quantify the dynamics of threatened risk of species under climatic change. Finally, I evaluated how climate and space could contribute species' occupancy using variance partitioning of occupancy with partial ordination methods.

MATERIALS AND METHODS

Studied Taxa

I considered the distribution of 93 endemic reptiles found in China as the case study. Biodiversity and distributional patterns of amphibians of China has been well studied in previous studies (Chen and Bi 2007; Qian et al. 2007; Xie et al. 2007; Fei et al. 2010). However, the response of amphibians against climatic change has little been studied. Given that the global rapid decline and environmental sensitivity of amphibians (Wake 1991, 2007), evaluation of population change of endemic taxa under future climatic scenarios would throw new insights into amphibian conservation and protection.

Prediction of Species Potential Distribution under Current and Future Climatic Scenarios

I utilized MaxEnt program (Phillips et al. 2004, 2006; Elith et al. 2011) to perform species distributional modeling. MaxEnt is a presence-only modeling program and has been demonstrated to be powerful and effective (Elith et al. 2006, 2011). The current potential distribution was carried out by using current climatic layers as the input, while the future potential distribution (2050 year) was carried out by using future climatic layers as the input. The model evaluation was measured using AUC curve and cross validation.

The climatic layers being considered included annual precipitation, mean temperature, maximum temperature, minimal temperature and evaporation. These environmental layers were obtained from the following URL: http://www.worldclim.org.

Estimation of Species Abundance Based on Fundamental Niches of Species

Similar to a previous empirical study (Figueiredo and Grelle 2009; Maclean et al. 2011), I utilized the negative Binomial distribution probability (He and Gaston 2000, 2007) to estimate species abundance using the occupancy information as follows,

$$\begin{cases} N = M_1 k[(1-m_1/M_1)^{-1/k} - 1] \\ N = M_2 k[(1-m_2/M_2)^{-1/k} - 1] \end{cases} \quad (1)$$

Where M_1 and M_2 are the total numbers of cells for map 1 and map 2, respectively, and m_1 and m_2 are the numbers of occupied cells of the maps. Here two maps described the same object (i.e., species occupancy) but with different spatial resolutions. Although aggregation parameter k is not fully scale-independent, its change across scales is believed to be negligible (He and Gaston 2007; Figueiredo and Grelle 2009).

Second, a non-aggregation model, which is derived from the random sampling of species on the given areas, is calculated as,

$$p = 1-(1-a/A)^N \quad (2)$$

Therefore, when ones know the probability of occurrence of species at the areas, the abundance of species could be estimated accordingly as,

$$N = \log(1-p)/\log(1-a/A) \quad (3)$$

In my study, all the three estimation methods of abundances were implemented and compared. I took into account the uncertainties of estimation. For aggregation method, the uncertainty was considered for the second coarser rescaled map, which could be at any spatial resolutions. As such, the estimation of aggregation parameter and the abundance of species should be varied accordingly. For non-aggregation method, the uncertainty was derived from the selection of occurrence probability p for estimating abundance in equation (3) because p could range between 0 and 1 (but excluding both terminal values). For both methods, I considered taking the mean and standard deviation for all parameters and species abundance to measure the estimation uncertainties.

Partitioning of Variance for Species' Occupancy

Partial CCA method (Legendre and Legendre 1998; Legendre et al. 2005) was utilized to partition the variance attributed to different processes of species' occupancy. Variance partitioning using ordination-based methods was believed to be superior to Mantel test-based method (Legendre et al. 2008). As such, the CCA method was implemented. I considered two typical components of variance, including space and climate. Further, for climatic variables, I divided them into three parts: precipitation, temperature and evaporation respectively to reveal their independent contributions to the variance of species' occupancy.

Estimation of Variance of Species Occupancy Change Explained by Climate Change

I implemented multiple regression analysis to investigate the relationship of species occupancy change predicted by ENM modeling and the change of climatic variables from current to future scenarios. The adjusted R2 is considered as the explained variance of species' occupancy change caused by climate change.

RESULTS

Estimation of Species Abundance at Current and Future Climate Scenarios

Most model AUC values for the species have very high values (approaching 1), appendix I showed all the results. As such, the predicted suitable ranges of endemic amphibians in China using MaxEnt program were statistically valid.

The estimation of abundances of the 93 endemic amphibians using both aggregation and non-aggregation methods under current and future climatic scenarios were not shown here but available upon request.

Estimation of Population Change under Climate Change

Based on the estimation of population of each species at current and future climate scenarios, it was suggested that almost half of the species would have a pattern of declining population at 2050 year. 57 species would have reduced populations, while 47 ones would increase their populations. It was found that the non-aggregation model ignoring aggregation of species could have a good power to predict species abundance.

Estimation of Range Shifting patterns of Endemic Amphibians of China

It was found that over 93 endemic amphibians, 43 species showed range contraction patterns, while 58 species had range expansion patterns by comparing species' occupancy information before and after climate change.

It was found that under climate change, 8 species would have northward range-shifting patterns, 3 species would have eastward range-shifting patterns, and 11 species would have westward range-shifting patterns while 9 species would tend to expand their ranges towards southern part of the territory.

Association between Range Contraction/Expansion Patterns and Population Declining/Growing Patterns

By utilizing the occupancy-abundance model, it was found that there was a close relationship between range sizes of species and their populations. When species tended to have narrower (or wider) ranges under future climatic scenario, they would usually have a population declining (or growing) trend.

Environmental Determinants of Species' Occupancy

Climate was the principal driver of species' occupancy (56% and 57% of total variance explained under climate and future climatic scenarios, respectively), while the contribution of space is minor (1% variance explained). However, the interaction between climate and space still contributed to a substantial part of the variance in species' occupancy (17% of total variance). Further, the simultaneous interaction among temperature, precipitation and evaporation was the major factor underpinning the importance of climate on structuring species occupancy (46% of the total variance). The pure contribution of each independent climatic component accounted for no more than 10% of the total variance.

The Relationship between Species Occupancy Change and Climatic Change

Through multiple regression analysis, it was found that across 93 endemic amphibians, the occupancy probability change of 41 species over grid cells of China were strongly associated with the change of climatic variables (the explained variance by climate change was over 0.1, as measured by adjusted R^2). Among the species, 8 ones were highly correlated with the change of climate change (adjusted $R^2>0.3$). For example, species 10 was strongly correlated with climate change. When checking each explanatory climatic factor, it was found that both the change of temperature-relevant variables and evaporation could negatively affect the occupancy probability of the species. However, the role of precipitation was unclear. Increasing temperature and evaporation was predicted to reduce the occupancy of the species.

DISCUSSION

Range Shifting Patterns

It was found that future climatic warning may increase ecological opportunity for amphibians to increase their colonization successes. In China, most western part of the country has a relatively arid and dry condition. As such, under future climatic conditions, it was found that climate change would increase the precipitation of the areas.

By evaluating the environmental data, it was further found that precipitation would increase by 0.19%, minimal temperature 25.9%, maximal temperature 0.8%, and mean temperature 1.58% in 2050 year. In contrast, the annual evaporation would decrease by 0.21% in 2050 year.

Population Change Patterns

It was found that 46 species showed a population declining pattern in 2050 year. As such, climate warning seemed to open new ecological opportunity for some endemic species to better survive and lead to increasing population growth.

Non-aggregation estimation of abundance (equation 3) has been applied to assess species threatened status (He 2012). In the current study, the model was applied to evaluate species population change under climate change.

Interestingly, it was found that some species which have been identified as Critically Endangered species by IUCN would have the changes to grow their populations when climate is becoming warning. In contrast, some other species listed as Vulnerable in IUCN database would further reduce their populations in 2050 year. Therefore, for setting up approximate conservation strategies, these predicted population-declining species should be at the top priority of consideration.

A Comparison of Two Methods for Estimating Species Abundance from Occupancy Information

Surprisingly, it was found that there was a very low congruence of species abundance estimation using He-Gaston aggregation and non-aggregation methods. The estimated abundance data using both methods showed a dispersion pattern generally. The discrepancy should be derived from the estimation methods. For aggregation model, there was a positive occupancy-abundance relationship. In contrast, it was observed that there was a negative relationship between occupancy and estimated abundance derived from the non-aggregation model. The basic reason for this phenomenon should be due to the estimation equation (3). When the occupancy of species is increased, when the probability of occurrence of species predicted by ENM method is assumed to be fixed or change a little bit, the estimated abundance should decrease accordingly. In contrast, the estimation of abundance based on equation (1) was independent from the prediction of occurrence probability aroused from ENM modeling. As such, the positive relationship could be observed. In conclusion, the fundamental reason of observing negative occupancy-abundance relationships for endemic amphibians in the present study is due to the statistical method used for estimating species abundance.

Factors Determining Species' Occupancy

Based on the partial CCA analysis, it was suggested that climate played a central role determining species occupancy status. Over half of the total variance could be explained by climatic variables.

Interestingly, it was found that temperature, precipitation and evaporation played an equally important role to determine species' occupancy, as evidenced by another partial CCA analysis, which took climate variables only into considerations and divided them into three parts (Figure 1). Further, the synergy effect occupied over 45% percent of total variance from species' occurrence data. As such, the interaction of different climatic variables should be of most importance to determine the presence/absence probabilities of endemic amphibians. The synergy of temperature and precipitation on structuring regional biodiversity of China has been observed in previous studies as well (Qian et al. 2007; Qian 2010).

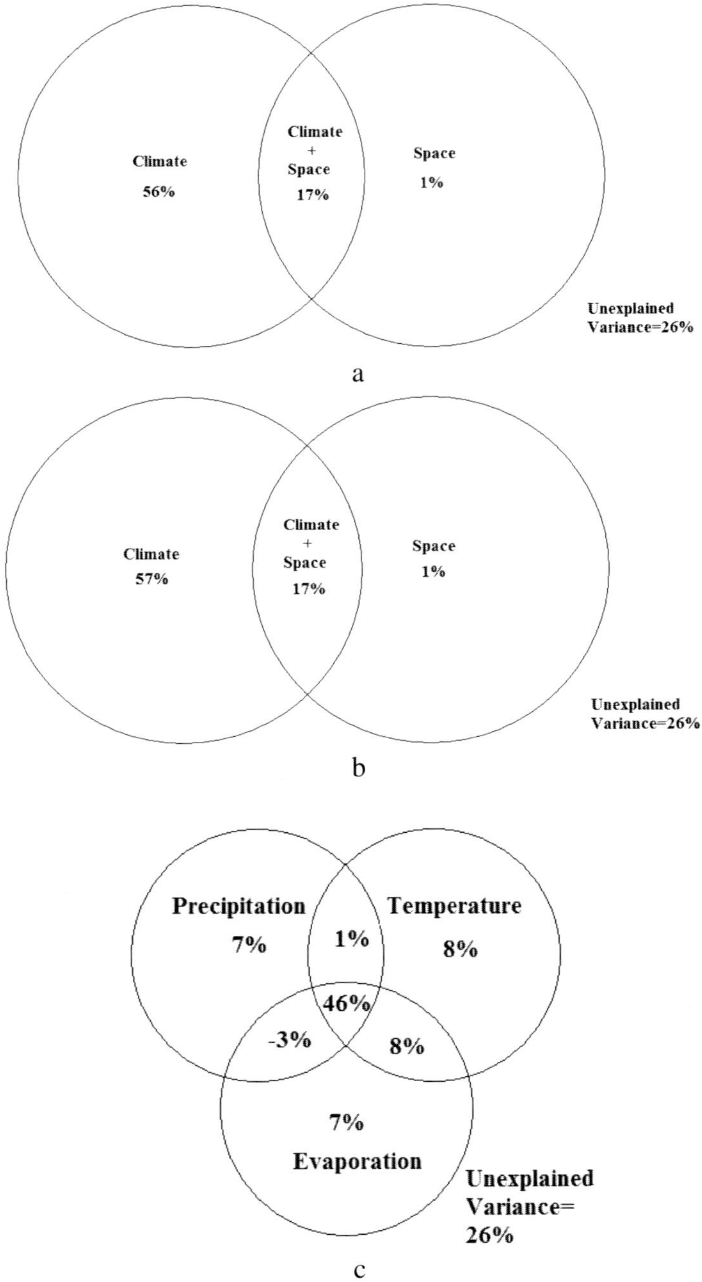

Figure 1. Variation partitioning of species' occupancy attributed to space and climate under current and future climate scenarios: a) current climate; b) future climate.

In light with these results, it was suggested that spatial autocorrelation might not be important in influencing species' occupancy, contradictory to the above-mentioned prediction. In contrast, climatic heterogeneity was very deterministic to affect species' presence.

REFERENCES

Blackburn T, Cassey P, Gaston K (2006) Variations on a theme: sources of heterogeneity in the form of the interspecific relationship between abundance and distribution. *J. Anim. Ecol.* 75:1426–1439.

Chen Y, Bi J (2007) Biogeography and hotspots of amphibian species of China: implications to reserve selection and conservation. *Curr. Sci.* 92:480–489.

Dormann C (2007) Effects of incorporating spatial autocoorrelation into the analysis of species distribution data. *Glob. Ecol. Biogeogr.* 16:129–138.

Elith J, Graham C, Aderson R,. (2006) Novel methods improve prediction of species' distributions from occurrence data. *Ecography* (Cop.). 29:129–151.

Elith J, Phillips S, Hastie T, Dudik M, Chee Y, Yates C (2011) A statistical explanation of MaxTent for ecologists. *Divers. Distrib.* 17:43–57.

Fei L, Ye C, Jiang J (2010) Progress and prospects for studies on Chinese amphibians. *Asian Herpetol. Res.* 1:64–85.

Figueiredo M, Grelle C (2009) Predicting global abundance of a threatened species from its occurrence: implications for conservation planning. *Divers. Distrib.* 15:117–121.

Frisk M, Duplisea D, Trenkel V (2011) Exploring the abundance-occupancy relationships for the Georges Bank finfish and shellfish community from 1963 to 2006. *Ecol. Appl.* 21:227–240.

Gaston K (1996) The multiple forms of the interspecific abundance–distribution relationship. *Oikos* 76:211–220.

Gaston KJ, Blackburn TM, Greenwood JJD, Gregory RD, Quinn RM, Lawton JH (2000) Abundance–occupancy relationships. *J. Appl. Ecol.* 37:39–59.

Hawlitschk O, Porch N, Hendrich L, Balke M (2011) Ecological Niche Modelling and nDNA Sequencing Support a New, Morphologically Cryptic Beetle Species Unveiled by DNA Barcoding. *PLoS One* 6:e16662.

He F (2012) Area-based assessment of extinction risk. *Ecology* 93:974–980.

He F, Gaston K (2000) Estimating species abundance form occurrence. *Am. Nat.* 156:553–559.

He F, Gaston K (2007) Estimating abundance from occurrence: an underdetermined problem. *Am. Nat.* 170:655–659.

He F, Gston K, Wu J (2002) On species occupancy-abundance models. *Ecoscience* 9:119–126.

Hui C, Boonzaaier C, Boyero L (2012) Estimating changes in species abundance from occupancy and aggregation. *Basic Appl. Ecol.* 13:169–177.

Hurlbert A, White E (2007) Ecological correlates of geographical range occupancy in North American birds. *Glob. Ecol. Biogeogr.* 16:764–773.

Legendre P (1993) Spatial autocorrelation: trouble or new paradigm? *Ecology* 74:1659–1673.

Legendre P, Borcard D, Peres-Neto P (2005) Analyzing beta diversity: Partitioning the spatial variation of community composition data. *Ecol. Monogr.* 75:435–450.

Legendre P, Borcard D, Peres-Neto P (2008) Analyzing or explaining beta diversity: Comment. *Ecology*: 3238–3244.

Legendre P, Legendre L (1998) Numerical ecology. *Elsevier Science BV*, Amsterdam.

Liu H, Wang W, Song G, Qu Y, Li S, Fjeldsa J, Lei F (2012) Interpreting the process behind endemism in China by integrating the phylogeography and ecological niche models of the Stachyridopsis ruficeps. *PLoS One* 7:e46761.

López-Cárdenas J, Gonzalez Bravo FE, Salazar Schettino PM, Gallaga Solorzano JC, Ramírez Barba E, Martinez Mendez J, Sánchez-Cordero V, Peterson AT, Ramsey JM (2005) Fine-scale predictions of distributions of Chagas disease vectors in the state of Guanajuato, Mexico. *J. Med. Entomol.* 42:1068–81.

Maclean I, Wilson R, Hassall M (2011) Predicting changes in the abundance of African wetland birds by incorporating abundance-occupancy relationships into habitat association models. *Divers. Distrib.* 17:480–490.

Medley K (2010) Niche shifts during the global invasion of the Asian tiger mosquito, Aedes albopictus Skuse (Culicidae), revealed by reciprocal distribution models. *Glob. Ecol. Biogeogr.* 19:122–133.

Peterson AT, Sánchez-Cordero V, Beard C Ben, Ramsey JM (2002) Ecologic niche modeling and potential reservoirs for Chagas disease, Mexico. *Emerg. Infect. Dis.* 8:662–7.

Peterson A, Vieglais D (2001) Predicting species invasions using ecological niche modeling: new approaches from bioinformatics attack a pressing problem. *Bio Science* 2 51:363–371.

Petitpierre B, Kueffer C, Broennimann O, Randin C, Daehler C, Guisan A (2012) Climatic niche shifts are rare among terrestrial plant invaders. Science (80-.). 335:1344–1348.

Phillips S, Anderson R, Schapire R (2006) Maximum entropy modeling of species geographic distributions. *Ecol. Modell.* 190:231–259.

Phillips S, Dudik M, Schapire R (2004) A maximum entropy approach to species distribution modeling. In: Brodley C, Cavazos J, Chan P, Jacob N, Kak A, Lippman R, Moss E, Pavlopoulou C, Precup D, Scheeff D, Stefanovic D, Utgoff P, Yurcik B (eds.) Proc. 21st Int. Conf. Mach. Learn. *ACM Press,* New York, pp. 655–662.

Qian H (2010) Environment-richness relationships for mammals, birds, reptiles and amphibians at global and regional scales. *Ecol. Res.* 25:629–637.

Qian H, Wang X, Wang S, Li Y (2007) Environmental determinants of amphibian and reptile species richness in China. *Ecography* (Cop.). 30:471–482.

Raxworthy C, Ingram C, Rabibisoa N, Pearson R (2007) Applications of ecological niche modeling for species delimitation: a review and empirical evaluation using day geckos (Phelsuma) from Madagascar. *Syst. Biol.* 56:907–923.

Reeves P, Richards C (2011) Species delimitation under the general lineage concept: an empirical example using wild North American hops (Cannabaceae: Humulus lupulus). *Syst. Biol.* 60:45–59.

Rodder D, Schmidtlein S, Veith M, Lotters S (2009) Alien invasive slider turtle in unpredicted habitat: a matter of niche shift or of predictors studied? *PLoS One* 4:e7843.

Soberon J, Peterson A (2011) Ecological niche shifts and environmental space anisotropy: a cautionary note. *Rev. Mex. Biodivers.* 82:1348–1355.

Urbina-Cardona J, Flores-Villela O (2010) Ecological-niche modeling and priortization of conservation-area networks for Mexican herpetofauna. *Conserv. Biol.* 24:1031–1041.

Venette R, Kriticos D, Magarey R, Koch F, Baker R, Worner S, Raboteaux N, McKenney D, Dobesberger E, Yemshanov D, Barro P, Hutchison M, Fowler G, Kalaris T, Pedlar J (2010) Pest risk maps for invasive alien species: a roadmap for improvement. *Bioscience* 60:349–362.

Wake D (1991) Declining amphibian populations. *Science* (80-.). 253:860.

Wake D (2007) Climate change implicated in amphibian and lizard declines. *Proc. Natl. Acad. Sci. United States Am.* 104:8201–8202.

Webb T, Freckleton R, Gaston K (2012) Characterizing abundance-occupancy relationships: there is no artefect. *Glob. Ecol. Biogeogr*. 21:952–957.

Wilson P (2011) The consequences of using different measures of mean abundance to characterize the abundance–occupancy relationship. *Glob. Ecol. Biogeogr.* 20:193–202.

Xie F, Lau M, Staurt S, Chanson J, Cox N, Fischman D (2007) Conservation needs of amphibians in China: a review. *Sci. China Ser. C Life Sci.* 50:265–276.

Chapter 10

'COEXIST': AN R PACKAGE FOR PERFORMING SPECIES COEXISTENCE MODELING AND ANALYSIS

ABSTRACT

This letter introduced a new R package *'coexist'* which can perform species coexistence simulation and analysis. The package was initially developed for understanding the role of different combinations of varying species growth rates, dispersal ability, habitat heterogeneity, and competition interaction on influencing species coexistence under fluctuating environment. Further versions could introduce more coexistence models under different ecological conditions. The source package is available at the Google Code (https://code.google) and R CRAN repository (http://cran.r-project).

Keywords: coexistence mechanism, species modeling, fluctuating environment

INTRODUCTION

Understanding biodiversity maintenance mechanisms is the principal objective in ecological studies (Bascompte et al. 2006; Chesson 2000; Salomon et al. 2010). Many theories have been proposed to reveal species coexistence mechanisms, two principal ones were identified as stabilizing and equalizing mechanisms (Cheeson 1994; Chesson 2000; Chesson and

Neuhauser 2002). Stabilizing mechanism supported niche theory while equalizing mechanism advocated for neutral theory.

There were a lot of theoretical work and numerical simulations for explaining species coexistence reasons. However, up to date, there is not a software or program available to perform species coexistence analysis. Also, the role of asymmetric dispersal and differentiation of species' biological traits under fluctuating environments has not been widely understood. To cope with these gaps, I developed an R (R Development Core Team 2013) package called *'coexist,'* which can perform flexible simulations of the outcomes of coexistence of two or multiple species under the conditions that species traits, like intrinsic growth rates, dispersal ability, biotic interaction degree, were varied. The environmental conditions for the species can be varied as well.

Limitation of the current version for the package is that there were many loops involved in the code without optimization. This situation will lead to intensive consumption of computation time, and the pre-configured model with two species and simulation time of 10000 steps will lead to 3 hours. However, if the readers just want to learn how to perform individual-based species coexistence simulation and modeling, the package should be valuable.

The resultant simulation outcomes could be read, saved and opened using the tailored functions in the package. Typically the files would be saved with a postfix "00yh.dat," which will be served as the sign for batch reading of all the files inside a folder. Therefore, if the users have many models for simulations, it is quite convenient to analyze these models simultaneously using the batch-handling commands in the packages.

Currently, the package also integrated several graphical functions as well. Coexistence contour plots for the combination of each pair of parameters can be generated in the form of heatmaps (different colors in the grids indicated different coexistence probabilities).

Figure 1 showed a heatmap for illustrating the outcomes generated from the linear combination of two parameters' value spaces: growth rates for two species (r1 and r2). The result was the outcome generated by the Model 1 in the following text.

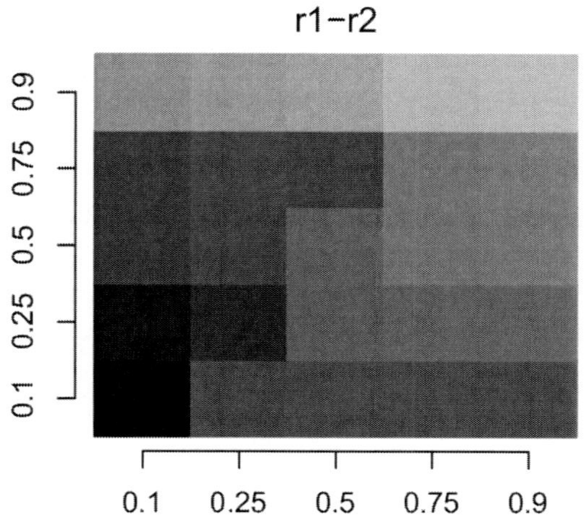

Figure 1. Coexistence probability by partialling out the influence of other parameters but keeping the effect from the parameter pair combination of the varying growth rates of two species model. Darker color indicated higher coexistence probability. Necessary model configuration for this simulation in "coexist" package was (1) habitat quality for species 1 is decreased from source to sink patches; while for species 2 is always in transition across the patches; (2) connectivity between the neighboring patches was decreased from the source to sink. It was derived from the Model 1.

DEMONSTRATION

Here is the procedure to generate a model configuration and perform analysis.

Firstly, in the package there is a simple function *sampletype* to help generate model names (in string form) with corresponding numbers (thus it should be easier without writing model names repeatedly). I can script the spatial pattern names as well.

```
#spatial patterns of a a single parameter in the simulations
#1 = decrease,2 = increase,3 = constant,4 = mosaiclow,5 = mosaichigh
modelnum=5
h<-vector()
h[1]<-"decrease"
h[2]<-"increase"
h[3]<-"constant"
h[4]<-"mosaiclow"
h[5]<-"mosaichigh"
```

It is noted that in the packages only spatial patterns were provided: "decrease" indicated that the parameter value is decreased from source to sink patches, while "increase" is opposite. "constant" demonstrated that the parameter is not influenced by the patch ordering. Last, "mosaiclow" is the one-by-one switching of the values (one of two values, either the high or low one, will be assigned to the parameter) between neighboring patches, but the first value assigned to the parameter (in the second patch adjacent to the source) is set to low. "mosaichigh" is almost the same, but the first value occurred in the second patch is set to the high one.

If I set a combination of spatial patterns for each parameter as,

t < -c(h[1],h[4],h[3],h[3],h[1],h[3],h[3])

After defining a sampled parameter value space as,

parspace < -c(.1,.25,.5,.75,.9) #serve as the searching points for each parameter in the interval [0-1]

Then it is ready to use for performing simulation. Before doing that, I should take a view about the functions for simulations. There are several versions of simulation functions.

sim.coarse() is for 2-species model. If I assume that both species' dispersal abilities were not different and marked by the iconicity of neighboring patches, then I have the following 7 paramters in order: habitat quality/preference for species 1 and 2, species growth rates for 1 and 2, patch connectivity, competition coefficients for species 1 and 2.

fast.flexsim() is for multiple species model (species number >= 2), each species will have 4 parameters, because I considered that connectivity is not that important as species dispersal capabilities. Therefore, the parameter of habitat connectivity was replaced by two parameters for measuring dispersal ability of species 1 and 2 respectively. Therefore, in this flexible simulation, I would have parameter number = 4*species number.

Moreover, this function can relax the assumption that the source supply of species is constant across the simulation time. Instead, there three source supply patterns are available, which was generated from another function *flex.dispersal()*: "*cochange*" means that all species will have identical source supply for each time step; "*flexible*" means that all species will have their own source supply at each time step, but all followed a same normal distribution;

"*constant*" means that the species supply from the source is unchanged across time steps and identical among species.

flexsim() identical to the function *fast.flexsim()*, but the latter should be faster a little bit (but remember that there are so many loops in the package, all the simulations are not fast at all!).

sim.refinement(): if I am interested into exploring further about a specific pair parameter space while setting others unchanged. This function can help do the refinement for searching the parameters' values at a smaller interval. It is only for 7-parameter 2-species model, therefore is limited to use with *sim.coarse()*.

Now I can run the model with my configuration of spatial patterns for each parameter.

```
Model 1 <-sim.coarse (scale = 2, allee = 1, T = 500, prange = parspace,
type = t,
path = "c://test/out1")
```

In the function, parameter *allee* indicated the allee effect (how small the population will be wiped out). *scale* is scale factor used to weigh connectivity parameter (making species colonization rates smaller). *type* is the model configuration as above. *path* is the local disk path to save the simulation output data.

If I run multiple species model, I should firstly modify the spatial pattern vector to be the length of 2*species number and keep in mind that the parameters always follow the order as *(habitat quality for sp1,sp2...;growth rates for sp1,sp2...;dispersal rates for sp1,sp2...;competition rates for sp1,sp2...)*!

For example, if I still have 2 species (more species are OK, but the simulation time will be incredibly huge under R environment).

```
t<-c(h[2],h[2],h[3],h[3],h[3],h[3],h[3],h[3]) #indicate a simple model that habitat quality for the two species are better from source to sink patches
```

So, the model becomes,

```
Model2<-fast.flexsim(scale = 2,allee = 1,T = 400,parcomb = allcomb,
spnum = 2,sourcetype = "constant",type = t, path = "c://test/out2")
```

The parameters in *fast.flexsim* function are a bit different, which "*parcomb*" replaced "*prange.*" Because the the combinations of all parameters are calculated firstly prior to simulation, with the purpose to reduce simulation budget.

So, "parcomb" accepts a matrix, not a vector, which is the combination of parameter spaces. The generation of the matrix can use a function called make.parcomb() as follows,

```
parnum = 8
parv<-c(.1,.9) #paramter value space for a single paramter
allcomb<-make.parcomb(parv, parnum, path = "c://outcome/index.dat")
dim(allcomb)[1] # = 256
```

It will have 2^8 combinations = 256.

After simulation, I can plot the density plot as a heatmap for exploring the influence of pairwise parameter combination on influencing species coexistence.

The function *batch.mpaircomp()* can generate the mean and variance of density matrix for each pair of parameters. If I want to look at species coexistence status only varying one parameter, the function *batch.monepar()* can be used.

Plotting heatmaps and making pdf graphics have the functions *batch.pdf.pairpar()* and *batch.pdf.onepar()* respectively for pairwise parameter and single parameter cases.

For example, for the Model 2, I can have,

```
pairparlist<-batch.mpaircomp (Model2,coenum = 2,spnum = spnum, parameters = parv)
singleparlist<-batch.monepar (Model2,coenum = 2, island, spnum = spnum, parameters= parv)
batch.pdf.pairpar(pairparlist, pagesetup = c(2,2),path = "c://test/pairpars.coexist")
batch.pdf.onepar(singleparlist, pagesetup = c(2,2),path = "c://test/onepar.coexist")
```

Finally, by packing all the codes together, I can perform simulation of species coexistence under different parameter setting as follows,

```
#spatial patterns of a a single parameter in the simulations
#1 = decrease,2 = increase,3 = constant,4 = mosaiclow,5 = mosaichigh
library(coexist)
h<-vector()
h[1]<-"decrease"
h[2]<-"increase"
h[3]<-"constant"
h[4]<-"mosaiclow"
h[5]<-"mosaichigh"

#two species model without dispersal difference
t<-c(h[1],h[4],h[3],h[3],h[1],h[3],h[3])
parspace<-c(.1,.25,.5,.75,.9) #serve as the searching points for each parameter in the interval [0-1]

Model1<- sim.coarse(scale = 2,allee = 1,T = 1000,prange = parspace, type = t, path = "c://test/out1")

#multiple species model
parnum = 8
parv<-c(.1,.9) #paramter value space for a single paramter
allcomb<-make.parcomb(parv, parnum, path = "c://outcome/index.dat")
t<-c(h[2],h[2],h[3],h[3],h[3],h[3],h[3],h[3]) #indicate a simple model that habitat quality for the two species are better from source to sink patches
Model2<-fast.flexsim(scale = 2,allee = 1,T = 300,parcomb = allcomb, spnum = 2, sourcetype = "constant", type = t, path = "c://test/out2") #if T>1000, very slow

#heatmap plotting and saving
pairparlist<-batch.mpaircomp(Model2,coenum = 2,spnum = spnum, parameters = parv)
singleparlist<-batch.monepar(Model2,coenum = 2,island, spnum = spnum, parameters = parv)
batch.pdf.pairpar(pairparlist, pagesetup = c(2,2), path = "c://test/pairpars.coexist")
batch.pdf.onepar(singleparlist, pagesetup = c(2,2), path = "c://test/onepar.coexist")
```

References

Bascompte J, Jordano P, Olesen J (2006) Asymmetric coevolutionary netowrks facilitate biodiversity maintenance. *Science* (80-) 312:431–433.

Cheeson P (1994) Multispecies competition in variable environments. *Theor Popul Biol* 45:227–276.

Chesson P (2000) Mechanisms of maintence of species diversity. *Annu Rev Ecol Syst* 31:343–366.

Chesson P, Neuhauser C (2002) Intraspecific aggregation and species coexistence. *Trends Ecol Evol* 17:210–211.

R Development Core Team (2013) R: A Language and Environment for Statistical Computing, Vienna, Austria. ISBN 3-900051-07-0, URL http://www.R-project.org.

Salomon Y, Connolly SR, Bode L (2010) Effects of asymmetric dispersal on the coexistence of competing species. *Ecol Lett* 13:432–41. doi: 10.1111/j.1461-0248.2009.01436.x.

Chapter 11

FITNESS EQUIVALENCE PROMOTES SPECIES COEXISTENCE UNDER FLUCTUATING ENVIRONMENTS WITH SIMULTANEOUS COUPLING OF DISPERSAL AND COMPETITION

ABSTRACT

The present study is to study the coupling of dispersal and competition under fluctuating resource patterns on promoting species coexistence. By using a simple but flexible source-sink model, my results indicated that neutral process can generate high degree of coexistence across the patches through continuous supply of individuals for inferior species in superior species-dominated patches. This subsidizing mechanism thus surmounts the biodiversity loss caused by strong competition effects. However, niche process often generates the coexistence patterns in a small fraction of sink patches. Thus, my work is consistent with previous work that widely spreading species is a result of spatial dispersal limitation, while range-restricted species is caused by environmental filtering niche process. I thus propose a local neutrality-regional niche hypothesis to maintain biodiversity.

Keywords: biodiversity maintenance, coupling of ecological mechanisms, asymmetric processes, competition

Introduction

Understanding species persisting coexistence under fluctuating environments is one central and vital topic in ecological studies. However, it is not clear about the link between dispersal and coexistence under a source-sink system, although there is a lot of works trying to understand the role of landscape connectivity (e.g., Vuilleumier and Possingham, 2006). Moreover, there are many efforts focusing on studying the stability of equilibrium points in competition models (e.g., Cushing et al., 2004). However, the space elements on influencing species coexistence were still limited in tradition, but nowadays emphasized gradually in coexistence theory (Yu et al., 2004; Berkley et al., 2010).

It is a widely debated issue about the niche and neutrality on explaining biodiversity patterns since the past ten years (Rosindell et al., 2011). Niche process is hard to explain the mega-biodiversity patterns in tropical forests (Zhou and Zhang, 2006), driving the birth of neutral theory of biodiversity and biogeography (Hubbell, 2001). Fitness equivalence (Hubbell, 2006) is the core of neutral theory, for which "trophically similar species are demographically alike on a per capita basis." Thus, fitness equivalence assumed that the growth rates of both species should be identical. By modifying parameter space and forming different combinations of parameters, we are able to obtain and compare neutrality and niche-driven coexistence outcomes, depending on whether the intrinsic growth rates of both species are equal or not.

Dispersal is important to maintain species coexistence. Typically different dispersal patterns, like local dispersal (Snyder and Chesson, 2003), turbulent dispersal (Berkley et al., 2010) or clumped dispersal (Potthoff et al., 2006) can facilitate species coexistence. Asymmetric dispersal, one specific form of these dispersal patterns, is thought to have profound impacts on species coexistence as well, and its role on species maintenance is still controversial. For instance, Salomon et al. (2010) found that if asymmetric dispersal was different among species, the coexistence is promoted even when lacking of environmental fluctuation. However, Vuilleumier and Possingham (2006) found that metapopulation is decreased at the double ratio compared to symmetric dispersal. They suggested that extinction risk is higher in asymmetric case. Further, Bode et al. (2008) suggested that asymmetric connectivity among patches will reduce population persistence. I implemented the spatially asymmetric dispersal in the present work because we have to use one patch serving as a source to offer new conspecific individuals to the system and look at the coexistence outcomes after the dispersal from the source patch to other sink patches and the competition within each patch.

Transitions in ecological systems are an interesting topic as they can provide ecologists ideas about the consequences of their experiments and inference of possible ecological mechanisms. So far, many literatures have contributed to analyze different facets of ecological transitions. For example, Chisholm and Filotas (2009) studied the transitions in two species models. Holland and DeAngelis (2009; 2010) used consumer-resource theory to predict dynamic switching among different patterns of species interaction. Multiple outcomes of species coexistence were found by varying the parameters in the equations. Six kinds of species interaction scenarios could emerge (Holland and DeAngelis 2009).

Spatially implicit assumption was often assumed in previous literature of coexistence theory to reduce complexity of analysis. However, in reality, spatially-fluctuated landscape variation and geometric constraints of dispersal in spatially explicit models often predicted different ecological outcomes and species interaction (e.g., Morozov and Petrovskii 2009). Therefore, spatial explicit models should be widely adopted to uncover new knowledge on species coexistence and transition patterns.

The simultaneous coupling of dispersal and competition on influencing species coexistence has seldom been studied as we know. The simultaneous responses of immigration and coexistence might be able to switch the alternative stable competition states. Thus, the prediction of species coexistence aroused might be more complicated than decoupling models. For example, using ant community as the experimental model, Yu et al. (2004) found out that the coupling of different processes, including resource partitioning, colonization, limited dispersal and competition, will influence coexistence, and the trade-off between long-distance flying ability of *Azteca* and superior fecundity of *Allomerus* can promote coexistence of both species.

Therefore, it is very interesting to build a clear theoretical framework to better understand the compromise effect of biodiversity gain caused by dispersal and biodiversity loss caused by competition, both of which can occur simultaneously, on promoting global coexistence across local island models. There are actually some theoretical work identifying that fast dispersal can overcome the competition. For example, Ngoc et al. (2010) have identified that under homogeneous landscape, fast dispersal can lead to inferior competitor dominance, for which originally superior competitor should occupy. However, it is still unclear if the reverse pattern can emerge again under heterogeneous habitats. Also, from biodiversity conservation perspective, we might expect that both inferior and superior competitors should be in harmonious coexistence across the landscape. These kinds of works are still limited and not available. Hence, my questions that will be addressed are as follows: what's the role of the source patch on changing the global patterns of coexistence when supplying individuals of species persistently to down-stream patches? Can neutral and niche processes have different or similar coexistence outcomes?

MATERIALS AND METHODS

Lotka-Volterra Species Competition Model

The simple Lotka-Volterra competition model, which is widely applied in assessing species coexistence (MacArthur 1970, Chesson 1990), is as follows,

$$\begin{aligned} \frac{dN_1}{dt} &= r_1 N_1 \left(1 - \frac{c_{11} N_1}{K_1} - \frac{c_{12} N_2}{K_1}\right) \\ \frac{dN_2}{dt} &= r_2 N_2 \left(1 - \frac{c_{22} N_2}{K_2} - \frac{c_{21} N_1}{K_2}\right) \end{aligned} \qquad (1)$$

Here, N_1 and N_2 are the population size of species 1 and 2 respectively. r_1 and r_2 are the growth rate of population for each species. c_{11} and c_{22} are the intraspecific competition coefficients for each species. c_{12} and c_{21} can be regarded as the coefficients of interspecific interaction, which can be asymmetrical (i.e., $c_{21} \neq c_{12}$). In particular, we set $c_{11} + c_{12} = 1$ and $c_{22} + c_{21} = 1$ respectively, for representing the total species interaction effect.

Patch Models

I use one-dimensional spatial scale to illustrate the model. In each simulation, we have N patches each of which is placed in a line. Species will disperse out from a source patch (the leftmost one); and then colonize the patches from left to right substantially (Figure 1). Each patch has two resources each of which is suited for one species, which can allow us to assess the impacts of resource patterns, dispersal ability and competition between species simultaneously.

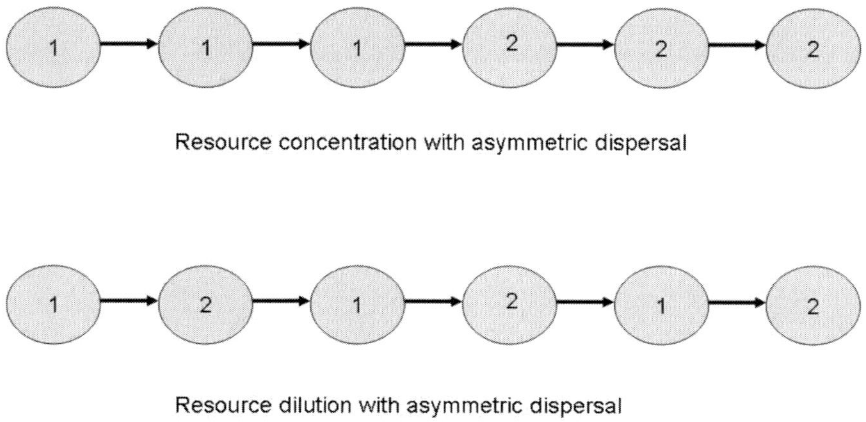

Figure 1. Source-sink patch models with different resource distributions. Asymmetric dispersal always happens from the source patch to sink patches. However, the competition coefficients, dispersal coefficients and resource partitioning across patches for each species can be varied.

Resource Fluctuation and Partition across Patches

Since this section, we have included the spatial elements and variations into the models to understand the impacts of resource fluctuation and dispersal variation on changing coexistence scenarios. The impact of resource variation is modeled by changing the capability of each patch for each species. Consistent with realistic worlds, a patch can be the one to 1) provide sufficiently preferred resources to both species; 2) provide only good resources to one species but not another, or 3) provide only bad (or non-preferred) resources to both species.

Distribution of resources across spatial landscapes should have two contrasting patterns basically. Resource concentration hypothesis predicts (Root 1973) that species (in particular, for herbivorous insects) will be more abundant in continuous dense favorite patches as they can easily catch the foods and have good condition of reproduction. In contrast, resource dilution hypothesis suggested the reversed pattern that species will have more abundant if resource is not concentrated or aggregated. In many empirical tests, the concentration hypothesis has been tested. For example, some researchers suggested that concentration effect was valid dependent of species (Grez and Gonzalez 1995). Both effects were observed in some works, for example, Otway et al. (2005) found both effects in the plots. Both effects may have different implications. For example, resource dilution can serve as a mechanism to reduce disease prevalence (Hall et al. 2009).

For reducing complexity, we only consider the three simple resource partition cases, which corresponded to the resource concentration and dilution cases outlined above:

1. homogeneous good/bad resource habitats: each patch have identical resource amount for the species, keeping the influence of habitats being constant across spatial scales.
2. good-bad resource one-time transition across habitats: the first half of patches is very preferred by both or one species, but the remaining ones only supply non-preferred resources to both or another species. It is a one-time shift of resources across the patches.
3. continuous switching of good-bad resource across the patches: the good and bad patches are assigned one-by-one in the space. Thus, when one species can encounter preferred resources in one patch, then by dispersal in the next neighboring patch it will encounter non-preferred resources, vice versa. This model is analogous to the resource dilution model in which the bad and good resources are maximally diluted in the continuous switching model.

A Spatiotemporal Competition Model

My model not only allow the change across spatial coordinates on the suitable resource amounts for each species, we can also allow that the other parameters, like intrinsic growth rates, competition varied across different patches. Equations (2) denote the refined Lotka-Volterra competition under spatial patch models. $f(d)$ represents that the coefficient is a function of locations (distance to the source patch, hereafters d represent the patch index from left to right). $f(d,t)$ denotes that the population size of each species is a function of both locations and time.

$$\frac{dN_1(d,t)}{dt} = r_1(d)N_1(d,t)(1 - \frac{c_{11}(d)N_1(d,t)}{K_1(d)} - \frac{c_{12}(d)N_2(d,t)}{K_1(d)})$$
$$\frac{dN_2(d,t)}{dt} = r_2(d)N_2(d,t)(1 - \frac{c_{22}(d)N_2(d,t)}{K_2(d)} - \frac{c_{21}(d)N_1(d,t)}{K_2(d)}) \quad (2)$$

Dispersal between Neighboring Patches

I considered the impacts of dispersal across patches on the resultant coexistence within each of the remaining patches (the first one was discarded which was served as a continuous source) for both species. Imaging that we have a total number of k patches in the model, then

$$\vec{N}_1(\vec{d},t+1) = M\vec{N}_1(\vec{d},t)$$
$$\vec{N}_2(\vec{d},t+1) = M\vec{N}_2(\vec{d},t) \quad (3)$$

Here $\vec{N}_i(\vec{d},t) = (N_i(1,t), N_i(2,t), \cdots, N_i(k,t))^T$ is the vector of population size across patches for species i . T denotes the transpose. M represents the connectivity matrix among the patches. In initial simulation, we assumed that both species can have equal and sufficient dispersal ability to immigrate into new patches continuously (This assumption could be relaxed). Thus, the dispersal rates across different patches for both species can be varied according to either habitat conditions or not. As my patch model is a line form, thus, the M matrix can be a form as,

$$M = \begin{pmatrix} d_{1,1} & d_{1,2} & & 0 & \\ & d_{2,2} & d_{2,3} & & \\ & & \cdots & & \\ & & & d_{k-1,k-1} & d_{k-1,k} \\ & 0 & & & 1 \end{pmatrix} \quad (4)$$

It is a Markov transition matrix, thus, we set $\sum_{j=i}^{i+1} d_{i,j} = 1$ for each row.

So, the total abundance for each species across the sites after reaching equilibrium (excluding those derived from source patch, which are always maintained to be a constant), can be written as,

$$\begin{aligned} N_1 &= \lim_{t\to\infty} \sum_{d\neq 1} N_1(d,t) \\ N_2 &= \lim_{t\to\infty} \sum_{d\neq 1} N_2(d,t) \end{aligned} \tag{5}$$

As seen, the coupling of equations (2) and (3) determines the equilibrium species abundance for each patch, which is dependent on the species abundance established in the previous patch, the immigration matrix among the patches, and the competition between both species.

Because of the coupling of simultaneous dispersal and competition, it is hard to get the equilibrium solution for both species coexistence across all patches. So, we performed simulation by putting both species in the first patch to see the resultant coexistence scenarios after equilibrium.

Simulation Schemes

At the source patch, the species abundance is always to be kept 1000. This initialization value for both species at the first patch is an equilibrium point (either stable or not) for the Lotka-Volterra competition system.

I now set the model with seven parameters, four competition coefficients (two for intraspecific competition and two for interspecific competition of the species), one dispersal parameter across the patches, and two intrinsic per capita growth rates of both species. As the sum of interspecific and intraspecific competition parameter values is 1, as long as either intraspecific or interspecific coefficient is determined, the other one can be determined as well. Thus, actually my system involved five effective parameters.

Table 1. 30 models for showing different combinations of resource partitioning, competition status, growth rates and dispersal across patches. *Model 20 has a different source species supply mechanism different from others (which are set to be constant), which is fluctuating across time by a normal distribution (mean = 1000, sd = 500)

Model number	Status of biotic and abiotic variables (assuming that species dispersal rates always high, that is, they are equivalent in dispersal limitation, so Models 1-19 used connectivity probabilities to indicate species dispersal across patches) c: constant across patches; d: decrease from source to sink patches; i: increase from source to sink patches; h: one-by-one switching with the variable condition in the first patch is high; l: one-by-one switching with the variable condition in the first patch is low					
	Habitat for species 1	Habitat for species 2	Growth rate for species 1	Growth rate for species 2	Connectivity among patches	Intra-Specific competition rate for species 1	Intra-Specific competition rate for species 2
1	d	d	c	c	c	c	c
2	d	d	c	c	d	c	c
3	l	l	c	c	c	c	c
4	l	l	c	c	l	c	c
5	d	l	c	c	c	c	c
6	d	l	c	c	l	c	c
7	d	l	c	c	d	c	c
8	d	h	c	c	c	c	c
9	d	h	c	c	h	c	c
10	d	h	c	c	d	c	c
11	i	i	c	c	c	c	c
12	i	i	c	c	i	c	c
13	i	l	c	c	c	c	c
14	i	l	c	c	l	c	c
15	i	l	c	c	i	c	c
16	d	d	c	c	c	d	d
17	d	d	d	d	c	c	c
18	d	d	c	c	c	c	d
19	d	d	d	c	c	c	c

Table 1. (Continued)

Model number	Additional models with the independent change of dispersal rates for both species, here, dispersal rates for both species are different, and connectivity across patches is assumed to be too low)						
	Habitat for species 1	Habitat for species 2	Growth rate for species 1	Growth rate for species 2	Dispersal rate for species 1	Dispersal rate for species 2	Intra-Specific competition rate for species 1	Intra-Specific competition rate for species 2
20*	d	d	c	c	c	d	c	c
21	d	i	c	c	i	d	c	c
22	d	d	c	c	c	c	c	c
23	c	c	c	c	c	c	c	c
24	d	h	c	c	d	l	c	c
25	l	i	c	c	h	d	c	c
26	h	i	l	d	l	d	l	d
27	l	i	h	d	h	d	h	d
28	d	i	i	d	i	d	i	d
29	d	c	i	c	i	c	i	c
30	d	c	i	c	i	c	i	c

For each parameter, we allowed it vary from 0.1 to 0.9, but for reducing simulation time and complexity, we only consider five specific parameter value points $(0.1, 0.25, 0.5, 0.75, 0.9)$. Thus, in a total, my models will examine $5^5 = 3125$ combinations of parameter vectors.

Because we used a simulation to reach equilibrium of the whole patch system by setting a random initial values, for each model with a specific parameter combination, we allow the system to run up to T = 1000 time step, which is enough to reach an equilibrium of the system.

I set up 30 scenarios to predict the impacts of different combinations of parameters on changing coexistence patterns. Table 1 summarized the configuration for each model. Here we introduce the features of some models. Models 1-6 illustrated the different resource partitioning scenarios, in which both species' resources co-occur at each patch. Models 16 and 19 allowed that other biotic interaction parameters, including intrinsic growth rate (16) and inter-/intra-specific competition coefficients (19), vary according to different patches and resource conditions.

Models 20-30 allowed the independent change of dispersal rates for both species, of which the dispersal rates for both species could be different. In this way, physical connectivity of patches was largely influenced by the dispersal ability of independent species. Therefore, in these models, there were 8 parameters (4 for each species). In particular, Models 26 and 27 allowed that the each of all pairs of parameters varied for each species. Model 20 has a different source species supply mechanism different from others (which are always set to be constant). The source supply was fluctuating across time by a normal distribution (mean = 1000, sd = 500).

RESULTS

Coexistence Caused by Neutrality versus Niche Processes

Figure 2 illustrated that how neutral process (fitness equivalence) and niche process differentiate the resultant coexistence outcomes. Interestingly, for neutral cases, the frequency of full coexistence is always higher than those attributed by niche process. Thus, we argued that high coexistence

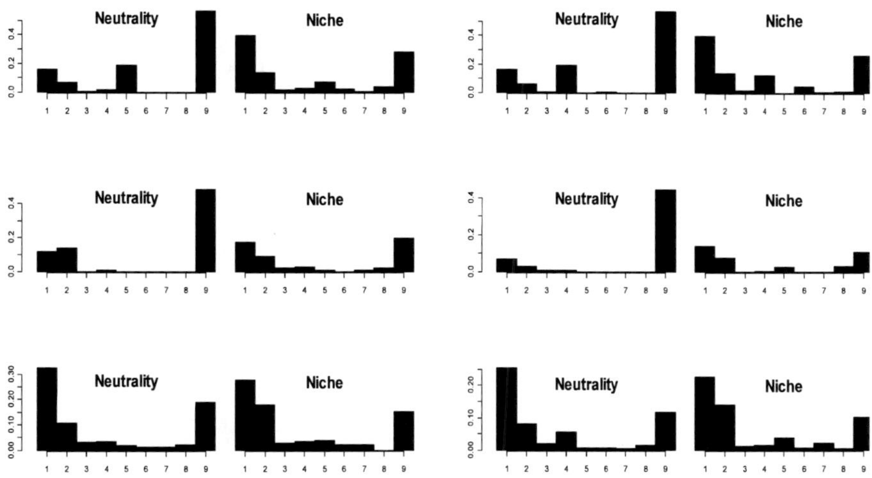

Figure 2. Comparison of coexistence outcomes derived from fitness equivalence and inequivalence parameter combinations in six simulation scenarios. For full coexistence cases (coexisting patch number = 9), it is clear that neutrality always outcompete niche. However, for partial coexistence cases, especially the coexisting patches are small (patch number = 1 or 2), niche models will always generate a higher amount of partial coexistence patterns.

Analysis on the Change of a Single Parameter on Influencing Species Coexistence

Figure 3 showed the coexistence density plots when each parameter changes but controlling out the influence of other parameters. The results are straightforward: increasing immigration rates between neighboring patches solely or increasing intraspecific competition coefficient for each species solely will always lead to a high degree of coexistence across the patches (Figure 3). Increasing intrinsic growth rate solely of one species will don't have any impacts on changing resultant coexistence patterns. Thus, negative density dependence mechanism is clear to promote global coexistence of species.

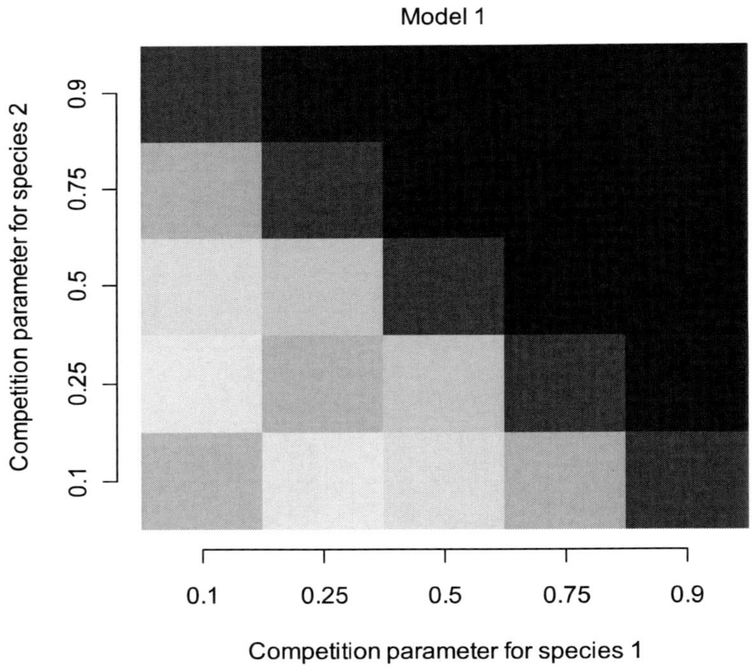

Figure 3. Competition density plot for different combinations of the intraspecific competition parameter pairs for two species. Each density plot is symmetric as expected. Gray colors from light to heavy indicate the average coexistence patch number ranging from low to high.

However, there are several interesting patterns contrasting my intuitions as well. First, in the case of one-time transition of habitats, although increasing dispersal rates will make the full coexistence become easy to realize, the medium-degree partial coexistence scenarios (coexisting patch number = 5) become disappear (Figure 3). This patch number is the transition patch where the habitat type was switched from one to another. However, there are still a lot of low coexistence cases in high dispersal rates. Thus, all the medium-degree partial coexistence cases which harbors transition patches will be converted to full coexistence much easier. Second, in the case of self-limited competition coefficient of one species, when the intraspecific competition coefficient is low, there are a lot of low partial coexistence cases (coexisting patch number = 1 or 2).

Invasibility Analysis of a Pair of Parameters on Influencing Species Coexistence

Pairwise comparison of coexistence status also confirmed the results from single parameter analysis. Typically, high dispersal rates plus high intraspecific competition will lead to full coexistence scenarios usually.

Coexistence Caused by Source-Sink Dispersal Process

For many cases, species are possible to coexist in some patches, where should be dominated by only superior species. The reason of coexistence is the continuous individual supply of inferior species from the source patch, which can sustain the species loss due to severe competition.

Discussion

A Local Neutrality-Regional Niche Hypothesis of Coexistence

Under source-sink system, my model demonstrated that power of neutral processes on promoting coexistence of species across patches. The reason for the facilitation of coexistence is due to the unstable equilibrium points in which both species can coexist can be maintained easier under fitness equivalence cases (neutrality). As for niche process, in which the intrinsic per capital growth rates of both species are different, the fragile coexistence scenario of both species can be easily removed.

Of course, it is important to maintain the subsidizing mechanism for which individuals of inferior species are offered by a source patch, thus, my model can be seen as a local island model, for which individuals of species must be dispersed from the metacommunity pool. In such a sense, it seems that niche process can facilitate coexistence at local scale.

Although it is challenged that niche process at metacommunity scale will reach a singular scenario in which only one species is found after many time steps, which violated biodiversity maintenance principle. Thus, at a broad scale, we might expect that niche process is dominated to maintain long-term persistence of biodiversity.

Thus, we argued that local neutrality-regional niche hypothesis is an appropriate way to understand species coexistence at multi-dimensional scales. Of course, this hypothesis has been used to model and compare the prediction of niche and niche processes as well. For example, the species abundance ranking derived from the regional area caused by niche process has been modeled by the summation of different independent small areas where independent neutral process is functioning (Chisholm and Pacala, 2010, 2011).

An Explanation of Widely-Spreading and Range-Restricted Species

My source-sink dispersal spatial model can have implications to possibly explain the causes of common distributed species and range-narrow species. As showed in Figure 2, niche process always has many cases of coexistence, for which the total coexistence patches are no more than 4 out of 9 patches. In contrast, neutral process can have a high degree of coexistence where the inhabiting patches are 9 out of 9 patches. Thus, we found that neutral process actually can facilitate the diversity of common distributed patches.

Coexistence in Spatial Explicit Models

My study provided new insights into the coexistence scenarios under spatially explicit models. As a long-term tradition, previous works only focus on spatially implicit models, which the role of spatial elements are always overlooked.

For better understanding coexistence scenarios, we argued that it is necessary to adopt a spatial hierarchical classification to include different coexistence possibilities in the modeling. For example, a three-possibility spatial coexistence model can be as follows: 1) full coexistence (coexistence everywhere), 2) partial coexistence (coexistence only in a fraction of focused habitat), and 3) no coexistence (only one dominant species occur across the concerned habitat).

Coexistence and Dispersal

It is not biologically meaningful to conclude that species coexistence must strictly satisfy mathematical rules in Lokta-Volterra system. Instead, because there are a lot of complex processes that can function simultaneously in the community, superior species-dominated non-coexisting scenarios can be surmounted and coexistence can be promoted by other population-subsidizing mechanisms. Dispersal will complicate coexistence scenarios. For example, Ngoc et al. (2010) found that the outcome of coexistence can be reversed when fast dispersal is allowed in the homogenous landscaped. Further, my study demonstrated that continuous dispersal can greatly promote coexistence, especially for the patches which originally superior competitor should occupy the whole patch and inferior species must go extinct. The subsidizing mechanism caused from continuous dispersal can allow inferior species coexists in those areas with a small size of population.

My model should be a realistic scenario in real world for the simple reason that the dispersal rate is always set to be a much lower rates (10-times) compared to competition rates and growth rates. An additional simulation has the analogous results, in which dispersal events only happen rarely and randomly based on acceptance rate<.2 of the variate from a [0,1] uniform distribution run at each time step. At the meantime, competition is allowed to happen for every time step. Of course, the population size for inferior species in superior competitor-dominated patches becomes lower. But as long as the Allee effect is set to be a suitable threshold, no significant differences can observed.

Trade-Off among Resource Partitioning, Source-Sink Dispersal and Competition

It is clear that in my spatially explicit models, the trade-off between resource partitioning, source-sink dispersal and competition will promote global coexistence of the species. This kind of trade-off has been evidenced by previous experimental studies (e.g., Yu et al., 2004).

My model is very flexible, because all kinds of parameter combinations can be allowed. Compared to the previous work (Ngoc et al., 2010), my model can allow asymmetric competition, asymmetric dispersal and asymmetric resource partitioning between the focused species. Of course, symmetric cases are easy to set. Moreover, my model can allow the parameters vary across different patches (as the equation (2) can be a function of patch distance and time), for which the adaptation against new environments can be possible.

Open System versus Close System

My study improved and modified the recognition on unstable coexistence scenarios. It is always assumed that the unstable coexistence will be hard to maintain as a small perturbation will lead the differential systems move far away from unstable equilibrium points. However, we found out that the role of dispersal is important to maintain long-term unstable coexistence scenarios. As long as the dispersal rates versus competition rates are not too low, unstable coexistence scenarios can be maintained continuously and not affected by perturbation and disturbance.

Implications for Biodiversity Conservation

Nowadays asymmetric dispersal has been recognized and assessed when performing conservation priority planning (Beger et al. 2010). My study further enhanced the knowledge of asymmetric dispersal on promoting global coexistence of species.

It is a time to better adopt corridor hotspots, but not biodiversity hotspots to conserve local, regional and global dispersal across habitats. These corridors actually can greatly promote and maintain unstable coexistence scenarios by offering the compensation of species extinction due to competition exclusions within habitats.

One practical manner to recognize corridor hotspots should be those areas that are nearby the boundary edges of different habitats. However, it is noted that important corridor hotspots don't equal to edges of habitats. As showed in my simulation, the sites which can allow unstable coexistence of species are adjacent to but not the very edge sites between two habitats.

Density Dependence Regulation and Dispersal

Clearly my models illustrated that dispersal can overcome the limits of density dependence regulations. Thus, species don't have to limit themselves greater than other closed species in dispersal-regulated open systems.

Because dispersal rate is always limited to be largely smaller than competition rates (20-fold in most models). Thus, we can expect that the role of dispersal is very vital to open more coexistence opportunities when dispersal rates can be increased or long-distance dispersal events are accelerated.

FURTHER PERSPECTIVES

My simple but complete simulation demonstrated the transitional consequences of species coexistence across patches by switching from one stable equilibrium state to another stable equilibrium state. The limit of the work is only focused on the most common form of species interaction-competition. For potential theoretical simulation framework, it is of great value to consider the impact of other types of species interactions, for example, mutualism, amensalism, predation and so on (Holland and DeAngelis 2009).

CONCLUSION

My study highlights again the unique power of neutral process to promote coexistence of species at multiple patch models under the conditions of fluctuating resource partitioning, dispersal, and competition interactions. Pure neutral process can explain the highest degree of coexistence by co-occupying all patches. Nearly neutral model and niche model, however, explain why there are so many rare species with restricted distributional range across the habitats.

REFERENCES

Beger M, Linke S, Watts M, Game E, Treml E, Ball I, Possingham HP (2010) Incorporating asymmetric connectivity into spatial decision making for conservation. *Conservation Letters,* 3, 359-368.

Berkley HA, Kendall BE, Mitarai S, Siegel DA (2010) Turbulent dispersal promotes species coexistence. *Ecology Letters,* 13, 360-371.

Bode M, Burrage K, Possingham HP (2008) Using complex network metrices to predict the persistence of metapopulations with asymmetric connectivity patterns. *Ecological Modelling,* 214, 201-209.

Chesson P (1990) MacArthur's consumer-resource model. *Theoretical population biology*, 37, 26-38.

Chisholm RA, Filotas E (2009) Critical slowing down as an indicator of transitions in two-species models. *Journal of Theoretical Biology,* 257, 142-149.

Chisholm RA, Pacala SW. 2010. Niche and neutral models predict asymptotically equivalent species abundance distributions in high-diversity ecological communities. *Proceedings of the National Academy of Sciences* 107(36): 15821-15825.

Chisholm RA, Pacala SW. 2011. Independent species in independent niches behave neutrally: A response. *Oikos* 120(7): 964-965.

Cushing JM, Levarge S, Chitnis N, Henson SM (2004) Some discrete competition models and the competitive exclusion principle. *Journal of Difference Equations and Applications,* 10, 1139-1151.

Grez AA, Gonzalez RH (1995) Resource concentration hypothesis: effect of host plant patch size on density of herbivorous insects. *Oecologia,* 103, 471-474.

Hall SR, Becker CR, Simonis JL, Duffy MA, Tessier AJ, Caceres CE (2009) Friendly competition: evidence for a dilution effect among competitors in a planktonic host-parasite system. *Ecology,* 90, 791-801.

Holland, J.N. and D.L. DeAngelis. 2010. A consumer-resource approach to the density-dependent population dynamics of mutualism. Ecology 91:1286-1295. Holland JN, DeAngelis DL (2009) Consumer-resource theory predicts dynamic transitions between outcomes of interspecific interactions. *Ecology Letters*, 12, 1357-1366.

Hubbell SP (2001) The unified neutral theory of biodiversity and biogeography. *Princeton University Press,* Princeton.

Hubbell SP (2006) Neutral theory and the evolution of ecological equivalence. *Ecology*, 87, 1387-1398.

Morozov A, Petrovskii S (2009) Excitable population dynamics, biological control failure, and spatiotemporal pattern formation in a model ecosystem. *Bulletin of Mathematical Biology*, 71, 863-887.

Ngoc DN, Parra RB, Zavala MA, Auger P (2010) Competition and species coexistence in a metapopulation model: can fast dispersal reverse the outcome of competition in a homogeneous environment? *Journal of Theoretical Biology,* 266, 256-263.

Otway SJ, Hector A, Lawton JH (2005) Resource dilution effects on specialist insect herbivores in a grassland biodiversity experiment. *Journal of Animal Ecology,* 74, 234-240.

Potthoff M, Johst K, Gutt J, Wissel C (2006) Clumped dispersal and species coexistence. *Ecological Modelling,* 198, 247-254.

Root RB (1973) Organization of a plant-arthropod association in simple and diverse habitats: the fauna of collards (Brassica oleracea). *Ecological Monographs*, 43, 95-124.

Rosindell J, Hubbell SP, Etienne R (2011) The unified neutral theory of biodiversity and biogeography at age ten. *Trends in Ecology and Evolution,* 26, 340-348.

Salomon Y, Connolly SR, Bode L (2010) Effects of asymmetric dispersal on the coexistence of competing species. *Ecology Letters,* 13, 432-441.

Snyder RE, Chesson P (2003) Local dispersal can facilitate coexistence in the presence of permanent spatial heterogeneity. *Ecology Letters*, 6, 301–309.

Vuilleumier S, Possingham HP (2006) Dose colonization asymmetry matter in metapopulations? *Proceedings of the Royal Society B: Biological Sciences,* 273, 1637-1642.

Yu DW, WiLSON hb, Frederickson ME, Palomino W, Colina RDL, Edwards DP (2004) Experimental demonstration of species coexistence enabled by dispersal limitation. *Journal of Animal Ecology,* 73, 1102-1114.

Zhou SR, Zhang DY (2006) Neutral theory in community ecology. *Journal of Plant Ecology,* 30, 8680-877.

PART 2: APPENDICES OF R CODES USED FOR COMPUTATION AND SIMULATION

APPENDIX 1: R CODE FOR FISHER TRANSFORM AND SIMULATION

```
#WRITTEN BY YOUHUA CHEN
#distributed without warranty

#Fisher's Z-transform
Fisher<-function(x,y,is.distrzFALSE)
{
if(is.dist==FALSE)
{
xv<-dist(x)
yv<-dist(y)
}else
{
xv<-x
yv<-y
}
#
r<-cor(xv,yv)
d<-length(xv)
w<-1/2*log((1+r)/(1-r))
z<-sqrt(d-3)*w
res<-c(r,z)
names(res)<-c("Rm","Z")
return(res)
}
```

```
##############################################################
##############################################################
##############################################################
#FISHER'S Z-TRANSFORM#########################################
#The rationale of using Fisher's Z-transform is to quantify the
#significance of Mantel statistic
#randomly simulate matrices
#six treatments
library(vegan)
#
cc<-c(30,50,100,200,300,500)
rr1<-c(20,40,50,150,250,450)
rr2<-c(40,60,150,250,350,550)
mm<-vector()
mprob<-vector()
ff<-vector()
fprob<-vector()
mat<-vector()
for(i in 1:6)
{
for(j in 1:500)
{
c1<-cc[i]
r1<-sample(rr1[i]:rr2[i],1)
r2<-sample(rr1[i]:rr2[i],1)
x<-matrix(rnorm(c1*r1), nrow=c1,ncol=r1)
y<-matrix(rnorm(c1*r2), nrow=c1,ncol=r2)
#
out<-mantel(dist(x), dist(y))
value<-c(out$statistic, out$signif)
out1<-Fisher(x,y)
#
mm<-c(mm,value[1])
mprob<-c(mprob,value[2])
ff<-c(ff,out1[1])
fprob<-c(fprob,out1[2])


mat<-cbind(mat,c(c1,r1,r2))
```

```
}#j
}#i
#
```

APPENDIX 2: R CODE FOR SIMULATING ASYMMETRIC DISPERSAL AND COEXISTENCE

```
#testing the proportion of coexistence scenarios
#driven by neutral process versus niche process
#WRITTEN BY YOUHUA CHEN
#2012/4/13
#distributed without warranty


##################################################################
###############Simple functions for two species model################
##################################################################

#initialization
spabundance<-function(island,abund=1000)
{
sabund<-vector()
length(sabund)<-island
sabund[1]=abund
sabund[2:island]=0
return(sabund)
}

#rate matrix for each species at each island
parsetting<-function(island,rate=1,scale=2,type="decrease")
```

```
{
parset<-vector()
length(parset)<-island
if(type=="decrease")
{
parset[1:as.integer(island/2)]=rate
parset[(as.integer(island/2)+1):island]=rate/scale
}
if(type=="increase")
{
parset[1:as.integer(island/2)]=rate/scale
parset[(as.integer(island/2)+1):island]=rate
}
if(type=="constant")
{
parset[1:island]=rate
}
if(type=="mosaiclow")
{
for(i in 1:island)
{
if(i%%2==0)
{
parset[i]=rate
}else
{
parset[i]=rate/scale
}
}#for i
}
if(type=="mosaichigh")
{
for(i in 1:island)
{
if(i%%2==0)
{
parset[i]=rate/scale
}else
{
```

```
parset[i]=rate
}
}#for i
}
return(parset)
}

#dispersal parameter matrix
dispvar<-function(island,rate=.5,scale=2,type="decrease")
{
dismat<-matrix(0,ncol=island,nrow=island)
if(type=="decrease")
{
for(i in 1:as.integer((island-1)/2))
{
dismat[i,i+1]=rate
}
for(i in (as.integer((island-1)/2)+1):(island-1))
{
dismat[i,i+1]=rate/scale
}
diag(dismat)<-1-rowSums(dismat)
}
if(type=="increase")#heterogeneous increasing dispersal
{
for(i in 1:as.integer((island-1)/2))
{
dismat[i,i+1]=rate/scale
}
for(i in (as.integer((island-1)/2)+1):(island-1))
{
dismat[i,i+1]=rate
}
diag(dismat)<-1-rowSums(dismat)
}
if(type=="constant")#homogenous dispersal
{
for(i in 1:(island-1))
{
```

```
dismat[i,i+1]=rate
}
diag(dismat)<-1-rowSums(dismat)
}
if(type=="mosaiclow")#homogenous dispersal
{
for(i in 1:(island-1))
{
if(i%%2==0)
{
dismat[i,i+1]=rate
}else
{
dismat[i,i+1]=rate/scale
}
}
diag(dismat)<-1-rowSums(dismat)
}
if(type=="mosaichigh")#homogenous dispersal
{
for(i in 1:(island-1))
{
if(i%%2==0)
{
dismat[i,i+1]=rate/scale
}else
{
dismat[i,i+1]=rate
}
}
diag(dismat)<-1-rowSums(dismat)
}
return(dismat)
}

#competition parameters matrix
compvar<-function(island,rate=.5,scale=2,type="constant")
{
```

```
comp<-matrix(0,ncol=island,nrow=2)
comp[1,]=parsetting(island,rate,scale,type)
comp[2,]=1-comp[1,]
return(comp)
}

#sample random type
#1=decrease,2=increase,3=constant,4=mosaiclow,5=mosaichigh
modelnum=5
sampletype<-function(lowbound=1,modelnum)
{
if(lowbound==modelnum)
{
r=lowbound
}else
{
r<-sample(lowbound:modelnum,1,replace=T)
}
if(r==1)
{
type="decrease"
}
if(r==2)
{
type="increase"
}
if(r==3)
{
type="constant"
}
if(r==4)
{
type="mosaiclow"
}
if(r==5)
{
type="mosaichigh"
}
return(type)
```

```
}


#################################
#build a detailed model
#RESOURCE
habitat1<-parsetting(island,good,scale=2,type="decrease")
habitat2<-parsetting(island,good,scale=2,type="decrease")
resource<-rbind(habitat1,habitat2)

#GROWTH RATE
grow1<-parsetting(island,rate=1,scale=2,type="decrease")
grow2<-parsetting(island,rate=1,scale=2,type="decrease")
grow<-rbind(grow1,grow2)

#DISPERSAL matrix
dismat<-dispvar(island,rate=.5,scale=2,type="decrease")

#COMPETITION matrix
comp1<-compvar(island,rate=.5,scale=2,type="decrease")
comp2<-compvar(island,rate=.6,scale=2,type="decrease")


#################################
#perform dispersal analysis
dispersal<-function(spvector,dismat,allee=1)
{
spvector<-spvector%*%dismat
spvector[1,1]=good #serve as a source
spvector[2,1]=good #serve as a source
spvector[which(spvector<allee)]=0
return(spvector)
}


#perform competition analysis
competition<-function(spvector,resource,comp1,comp2,grow,allee=1)
{
islandnum<-dim(spvector)[2]
```

```
for(i in 1:islandnum)
{
if(spvector[1,i]!=0 & spvector[2,i]!=0)
{
s1<-spvector[1,i]
s2<-spvector[2,i]
spvector[1,i]=s1+(1-comp1[1,i]*s1/resource[1,i]-
comp1[2,i]*s2/resource[1,i])*s1*grow[1,i]
spvector[2,i]=s2+(1-comp2[1,i]*s2/resource[2,i]-
comp2[2,i]*s1/resource[2,i])*s2*grow[2,i]
}
if(spvector[1,i]!=0 & spvector[2,i]==0)
{
s1<-spvector[1,i]
spvector[1,i]=s1+(1-comp1[1,i]*s1/resource[1,i])*s1*grow[1,i]
}
if(spvector[2,i]!=0 & spvector[1,i]==0)
{
s2<-spvector[2,i]
spvector[2,i]=s2+(1-comp2[1,i]*s2/resource[2,i])*s2*grow[2,i]
}
#check impossible coexistence
if(spvector[1,i]<allee)
{
spvector[1,i]=0
}
if(spvector[2,i]<allee) #must be a full species body and non-negative
{
spvector[2,i]=0
}
}#end for
return(spvector)
}#end competition function


#perform species-specific dispersal and fluctuating source analysis
#dismat is now a list
flex.dispersal<-function(spvector,dismat,allee=1,type="constant")
{
```

```
if(type=="constant")
{
spnum<-length(dismat)
for(i in 1:spnum)
{
spvector[i,]<-spvector[i,]%*%dismat[[i]]
spvector[i,1]=good
}
}
if(type=="flexible")
{
spnum<-length(dismat)
for(i in 1:spnum)
{
spvector[i,]<-spvector[i,]%*%dismat[[i]]
spvector[i,1]=rnorm(1,mean=good,sd=good/10) #fluctuating source
}
}
if(type=="cochange")
{
spnum<-length(dismat)
newresource<-rnorm(1,mean=good,sd=good/10)
for(i in 1:spnum)
{
spvector[i,]<-spvector[i,]%*%dismat[[i]]
spvector[i,1]=newresource #fluctuating source
}
}
spvector[which(spvector<allee)]=0
spvector[which(spvector<0)]=0
return(spvector)
}

#perform flexible competition analysis allowing multiple species
flex.competition<-function(spvector,resource,grow,comp,allee=1)
{
spnum<-dim(spvector)[1]
islandnum<-dim(spvector)[2]
for(i in 1:islandnum)
```

```
{
s<-spvector[,i]
for(sp in 1:spnum)
{
spvector[sp,i]=s[sp]+(1-comp[sp,i]*s[sp]/resource[sp,i]-(1-
comp[sp,i])*sum(s[-sp])/resource[sp,i])*s[sp]*grow[sp,i]
#check impossible coexistence
if(spvector[sp,i]<allee)#must be a full species body and non-negative
{
spvector[sp,i]=0
}
}
}#end for
return(spvector)
}#end competition function


##########################################
#batch analysis of the output scenarios

batch.coexistence<-function(out,island=island)
{
colist<-list()
scenarionum<-length(out)
length(colist)<-scenarionum
for(i in 1:scenarionum)
{
colist[[i]]<-sta.coexistence(out[[i]],island)
}
return(colist)
}

#batch anlaysis of multiple coexistence summary tables
#depend on sta.mcoexistence
batch.mcoexistence<-function(out,island=island,spnum=2)
{
colist<-list()
scenarionum<-length(out)
length(colist)<-scenarionum
```

```
for(i in 1:scenarionum)
{
colist[[i]]<-sta.mcoexistence(out[[i]],island=island,spnum=spnum)
}
return(colist)
}

#batch analysis of niche and neutral cases
#depend on function-batch.coexistence
batch.n2n<-function(colist)
{
resultlist<-list()
scenarionum=length(colist)
length(resultlist)<-scenarionum
for(i in 1:scenarionum)
{
resultlist[[i]]<-sta.fitness(colist[[i]])
}
return(resultlist)
}

#plot distribution of niche and neutral coexistence patterns
#depend on function-batch.n2n
plot.n2n<-function(resultlist,island=island,pagesetup=c(1,1),path=NULL)
{
prop<-list()
length(prop)<-length(resultlist)
scenarionum<-length(resultlist)
temp<-matrix(0,ncol=2,nrow=island-1)
for(i in 1:scenarionum)
{
temp[,1]=rev(resultlist[[i]][1:9,1]/resultlist[[i]][island,1])
temp[,2]=rev(resultlist[[i]][1:9,2]/resultlist[[i]][island,2])
prop[[i]]<-temp
}
if(length(path)!=0)
{
randnum<-runif(1)
pos<-unlist(gregexpr("/",path))
```

```
folder<-substr(path,1,pos[length(pos)]-1)
dir.create(folder,showWarnings=F)
filename=paste(path,"00yh",randnum,".dat",sep="")
}else
{
randnum<-runif(1)
dir.create(folder,showWarnings=F)
filename=paste(folder,"n2nbarplot",randnum,".dat",sep="")
}
pdf(filename)
par(mfrow=pagesetup)
for(i in 1:scenarionum)
{
nnplot<-barplot(prop[[i]],beside=T,axisnames=F,col=1) #don't have x-
axis
y<-max(prop[[i]])
text(nnplot[5,1],y-.05,"Neutrality",font=2,cex=2)
text(nnplot[5,2],y-.05,"Niche",font=2,cex=2)
axis(1,at=nnplot[,1],labels=c(1,2,3,4,5,6,7,8,9))
axis(1,at=nnplot[,2],labels=c(1,2,3,4,5,6,7,8,9))
#axis(1,at=(nnplot[dim(temp)[1],1]+nnplot[1,2])/2,labels=paste("Scenario
",i,sep="-"),lty=0)
mtext(paste("Model",i,sep="-"))
}
dev.off()
}#end function

#batch analysis to explore coexistence density for concerned two
parameters
#depend on function-batch.coexistence
batch.paircomp<-
function(coexistlist,parnum=parnum,parameters=parspace)
{
scenarionum<-length(coexistlist)
pairlist<-list()
length(pairlist)<-scenarionum
for(i in 1:scenarionum)
{
```

```
pairlist[[i]]<-
sta.paircomparison(coexistlist[[i]],parnum=parnum,parameters=parameters)
}
return(pairlist)
}

#multiple species case
#depend on function-the output of batch.mcoexistence
#coenum determines the coexisting number of species for each patch
batch.mpaircomp<-
function(coexistlist,coenum,parnum=parnum,spnum,parameters=parspace)
{
if(is.list(coexistlist))
{
scenarionum<-length(coexistlist)
pairlist<-list()
length(pairlist)<-scenarionum
for(i in 1:scenarionum)
{
pairlist[[i]]<-
sta.mpaircomparison(coexistlist[[i]],coenum,spnum=spnum,parameters=param
eters)
}
return(pairlist)
}
}

#batch analysis to explore coexistence density for a varying parameter
#depend on function-batch.coexistence
batch.onepar<-function(coexistlist,parameters=parspace)
{
scenarionum<-length(coexistlist)
pairlist<-list()
length(pairlist)<-scenarionum
for(i in 1:scenarionum)
{
pairlist[[i]]<-sta.comparison(coexistlist[[i]],parameters=parameters)
}
return(pairlist)
```

```
}

#batch analysis to explore multiple species coexistence density for a varying parameter
#depend on function-batch.coexistence
batch.monepar<-
function(coexistlist,coenum,island,spnum,parameters=parspace)
{
if(is.list(coexistlist))
{
scenarionum<-length(coexistlist)
pairlist<-list()
length(pairlist)<-scenarionum
for(i in 1:scenarionum)
{
pairlist[[i]]<-
sta.mcomparison(coexistlist[[i]],coenum,island,spnum,parameters=parameters
)
}
return(pairlist)
}
}

#batch analysis to plot matrix values for one parameter matrix
#depend on make.heatmap
batch.pdf.onepar<-function(parmatlist,pagesetup=c(2,2),path=NULL)
{
scenarionum<-length(parmatlist)
parnum<-length(parmatlist[[1]])
if(length(path)!=0)
{
randnum<-runif(1)
pos<-unlist(gregexpr("/",path))
folder<-substr(path,1,pos[length(pos)]-1)
dir.create(folder,showWarnings=F)
filename=paste(path,"00yh",randnum,".pdf",sep="")
}else
{
randnum<-runif(1)
```

```
dir.create(folder,showWarnings=F)
filename=paste(folder,"singleparameter",randnum,".pdf",sep="")
}
pdf(filename)
par(mfrow=pagesetup)
for(each in 1:parnum)
{
for(i in 1:scenarionum)
{
xname=paste("Model",i,sep="-")
title=names(parmatlist[[i]])[each]
t<-parmatlist[[i]][[each]]
t<-t[order(t[,1],decreasing=F),]
t<-t[-1,-1]
make.heatmap(t,xname=xname,xlab=c(0.1,0.25,0.5,0.75,0.9),ylab=c(1:9),title=title)
}#i
}#each
dev.off()
}

#batch analysis to plot matrix values for pairwise parameter matrix
#depend on make.heatmap
batch.pdf.pairpar<-function(parmatlist,pagesetup=c(2,2),path=NULL)
{
scenarionum<-length(parmatlist)
parnum<-length(parmatlist[[1]][[1]])
if(length(path)!=0)
{
randnum<-runif(1)
pos<-unlist(gregexpr("/",path))
folder<-substr(path,1,pos[length(pos)]-1)
dir.create(folder,showWarnings=F)
filename=paste(path,"00yh",randnum,".pdf",sep="")
}else
{
randnum<-runif(1)
dir.create(folder,showWarnings=F)
filename=paste(folder,"pairwiseparameters",randnum,".pdf",sep="")
```

```
}

pdf(filename)
par(mfrow=pagesetup)
for(each in 1:parnum)
{
for(i in 1:scenarionum)
{
xname=paste("Model",i,sep="-")
title=names(parmatlist[[i]][[1]])[each]
t<-parmatlist[[i]][[1]][[each]]
t<-t(t)
make.heatmap(t,xname=xname,xlab=c(0.1,0.25,0.5,0.75,0.9),ylab=c(0.1,0
.25,0.5,0.75,0.9),title=title)
}#i
}#each
dev.off()
}


#################################################################
##############IMPORTANT SIMULATION FUNCTIONS#############
#################################################################

#coarse parameter sampling based on parspace
sim.coarse<-
function(scale=2,dispersalscale=51,allee=1,T=1000,prange=parspace,type=typ
evector,path=NULL)
{
parnum=5
parlen<-length(prange)
outcome<-list()
length(outcome)<-parlen^parnum
outindex<-matrix(0,nrow=parlen^parnum,ncol=parnum)
colnames(outindex)<-c("r1","r2","dis","c11","c22")

habitat1<-parsetting(island,good,scale,type[1])
habitat2<-parsetting(island,good,scale,type[2])
resource<-rbind(habitat1,habitat2)
```

```
#begin simulation
count=0
outcomefile=filename.check(path)
for(i1 in 1:parlen)
{
for(i2 in 1:parlen)
{
for(i3 in 1:parlen)
{
for(i4 in 1:parlen)
{
for(i5 in 1:parlen)
{
grow1<-parsetting(island,rate=prange[i1],scale,type[3])
grow2<-parsetting(island,rate=prange[i2],scale,type[4])
grow<-rbind(grow1,grow2)
dismat<-dispvar(island,rate=prange[i3]/dispersalscale,scale,type[5])
comp1<-compvar(island,rate=prange[i4],scale,type[6])
comp2<-compvar(island,rate=prange[i5],scale,type[7])
spvector<-rbind(spabundance(island,1000),spabundance(island,1000))
count=count+1
outindex[count,]<-
c(prange[i1],prange[i2],prange[i3],prange[i4],prange[i5])
for(j in 1:T)
{
spvector<-competition(spvector,resource,comp1,comp2,grow,allee)
spvector<-dispersal(spvector,dismat,allee)
}
outcome[[count]]<-spvector
write.table(outcome[[count]],file=outcomefile,sep="\t",append=TRUE)
outcome[[count]]<-list(abund=outcome[[count]],pars=outindex[count,])
}#for i1
}
}
}
}#for i5
#write.table(outindex,file=outindexfile,sep="\t")
return(outcome)
}#end function
```

```
#refinement of a specific pair parameter space
sim.refinement<-function(pars=c(p1,p2),interval=20,scale=2,dispersal
scale=51,allee=1,T=1000,parnum=parnum,prange=parspace,type=typevector)
{
parlen<-rep(length(prange),1,parnum)
parspace1<-list()
length(parspace1)<-parnum
for(i in 1:parnum)
{
parspace1[[i]]<-prange
}
outcome<-list()
length(outcome)<-length(prange)^(parnum-2)*interval^2
outindex<-matrix(0,nrow=length(outcome),ncol=parnum)
colnames(outindex)<-c("r1","r2","dis","c11","c22")
p1range<-vector()
length(p1range)<-interval
for(i in 1:interval)
{p1range[i]<-.01+(.99-.01)/interval*i}
p2range<-p1range

#function for changing parameter space for a given pair
for(i in 1:length(prange))
{
if(i==p1)
{
parlen[i]=length(p1range)
parspace1[[i]]=p1range
}
if(i==p2)
{
parlen[i]=length(p2range)
parspace1[[i]]=p2range
}
}

habitat1<-parsetting(island,good,scale,type[1])
habitat2<-parsetting(island,good,scale,type[2])
resource<-rbind(habitat1,habitat2)
```

```
#begin simulation
count=0
randnum<-runif(1)
outcomefile=paste("e://outcome",randnum,".dat",sep="")
outindexfile=paste("e://outindex",randnum,".dat",sep="")
for(i1 in 1:parlen[1])
{
for(i2 in 1:parlen[2])
{
for(i3 in 1:parlen[3])
{
for(i4 in 1:parlen[4])
{
for(i5 in 1:parlen[5])
{
grow1<-parsetting(island,rate=parspace1[[1]][i1],scale,type[3])
grow2<-parsetting(island,rate=parspace1[[2]][i2],scale,type[4])
grow<-rbind(grow1,grow2)
dismat<-
dispvar(island,rate=parspace1[[3]][i3]/dispersalscale,scale,type[5])
comp1<-compvar(island,rate=parspace1[[4]][i4],scale,type[6])
comp2<-compvar(island,rate=parspace1[[5]][i5],scale,type[7])
spvector<-rbind(spabundance(island,1000),spabundance(island,1000))
count=count+1
outindex[count,]<-
c(parspace1[[1]][i1],parspace1[[2]][i2],parspace1[[3]][i3],parspace1[[4]][i4],p
arspace1[[5]][i5])
for(j in 1:T)
{
spvector<-competition(spvector,resource,comp1,comp2,grow,allee)
spvector<-dispersal(spvector,dismat,allee)
}
outcome[[count]]<-spvector
write.table(outcome[[count]],file=outcomefile,sep="\t",append=TRUE)
outcome[[count]]<-list(abund=outcome[[count]],pars=outindex[count,])
}#for i1
}
}
}
```

```
}#for i5
write.table(outindex,file=outindexfile,sep="\t")
return(outcome)
}#end function

#function for coarse parameter sampling based on parspace, can allow multiple species
flexsim<-
function(scale=2,dispersalscale=51,allee=1,T=1000,prange=parspace,spnum=2,sourcetype="constant",type=typevector,path=NULL)
{
#habitat, growth rate, dispersal rates and competition rates-four parameter set models
parnum=spnum*3
outcome<-list()
if(!is.list(prange))
{
length(outcome)<-length(prange)^parnum
}else
{
parlen<-vector()
length(parlen)<-length(prange)
for(i in 1:length(prange))
{
parlen[i]<-length(prange[[i]])
}
length(outcome)<-prod(parlen)
}
resource<-matrix(0,ncol=island,nrow=spnum)
grow<-resource
comp<-resource

for(i in 1:spnum)
{
resource[i,]<-parsetting(island,good,scale,type[i])
}

#begin simulation
outcomefile=filename.check(path)
```

```
    if(!is.list(prange))
    {
    parcomb<-comblist(prange,parnum)
    }
    if(is.list(prange))#list of parameters to save space
    {
    parcomb<-comblist2(prange,parnum)
    }
    comnum<-dim(parcomb)[1]
    colnames(parcomb)<-
c(paste("r",c(1:spnum),sep=""),paste("dis",c(1:spnum),sep=""),paste("com",c(
1:spnum),sep=""))

    #construct list for storing dispersal matrix
    dismat<-list()
    length(dismat)<-spnum

    for(each in 1:comnum)
    {
    typenum<-spnum
    for(i in 1:spnum)
    {
    grow[i,]<-parsetting(island,rate=parcomb[each,i],scale,type[typenum+i])
    }
    typenum<-spnum+typenum
    for(i in 1:spnum)
    {
    dismat[[i]]<-
dispvar(island,rate=parcomb[each,i+spnum]/dispersalscale,scale,type[typenum
+i])
    }
    typenum<-spnum+typenum
    for(i in 1:spnum)
    {
    comp[i,]<-
parsetting(island,rate=parcomb[each,i+spnum*2],scale,type[typenum+i])
    }#finish setup of parameters
    spvector<-rbind(spabundance(island,1000),spabundance(island,1000))
```

```
for(j in 1:T)
{
spvector<-flex.competition(spvector,resource,grow,comp,allee)
spvector<-flex.dispersal(spvector,dismat,allee,type=sourcetype)
}

outcome[[each]]<-spvector
write.table(outcome[[each]],file=outcomefile,sep="\t",append=TRUE)
outcome[[each]]<-list(abund=outcome[[each]],pars=parcomb[each,])

}#end parcomb search

return(outcome)
}#end function

#function for faster multiple-species simulation
#parcomb is matrix and store all possible combinations before simulation
and apply to all simulations.
fast.flexsim<-
function(scale=2,dispersalscale=51,allee=1,T=1000,parcombination=parcomb,
spnum=2,sourcetype="constant",type=typevector,path=NULL)
{
#habitat, growth rate, dispersal rates and competition rates-four parameter
set models
parnum=spnum*3
parcomb<-parcombination
outcome<-list()
length(outcome)<-dim(parcomb)[1]
resource<-matrix(0,ncol=island,nrow=spnum)
grow<-resource
comp<-resource
spvector<-resource

for(i in 1:spnum)
{
resource[i,]<-parsetting(island,good,scale,type[i])
}

outcomefile<-filename.check(path)
```

```
comnum<-dim(parcomb)[1]
colnames(parcomb)<-
c(paste("r",c(1:spnum),sep=""),paste("dis",c(1:spnum),sep=""),paste("com",c(
1:spnum),sep=""))

#construct list for storing dispersal matrix
dismat<-list()
length(dismat)<-spnum

for(each in 1:comnum)
{
typenum<-spnum
for(i in 1:spnum)
{
grow[i,]<-parsetting(island,rate=parcomb[each,i],scale,type[typenum+i])
}
typenum<-spnum+typenum
for(i in 1:spnum)
{
dismat[[i]]<-
dispvar(island,rate=parcomb[each,i+spnum]/dispersalscale,scale,type[typenum
+i])
}
typenum<-spnum+typenum
for(i in 1:spnum)
{
comp[i,]<-
parsetting(island,rate=parcomb[each,i+spnum*2],scale,type[typenum+i])
}#finish setup of parameters
for(i in 1:spnum)
{
spvector[i,]<-spabundance(island,1000)
}
#finish abundance setup
for(j in 1:T)
{
spvector<-flex.competition(spvector,resource,grow,comp,allee)
spvector<-flex.dispersal(spvector,dismat,allee,type=sourcetype)
}
```

```
outcome[[each]]<-spvector
write.table(outcome[[each]],file=outcomefile,sep="\t",append=TRUE)
outcome[[each]]<-list(abund=outcome[[each]],pars=parcomb[each,])

}#end parcomb search

return(outcome)
}#end function

#function for making the parameter space matrix
make.parcomb<-function(prange,parnum,path=NULL)
{
if(!is.list(prange))
{
parcomb<-comblist(prange,parnum)
}
if(is.list(prange))#list of parameters to save space
{
parcomb<-comblist2(prange,parnum)
}
if(length(path)!=0)
{
path=filename.check(path)
write.table(parcomb,path,sep="\t")
}
return(parcomb)
}#end function


#####################################################################
##################Post-statistic and handling functions################
#####################################################################

#check outcome and analysis

#function for checking the number of occupied islands/sites
sta.sitenum<-function(outcome)
{
sta1<-matrix(0,ncol=3,nrow=length(outcome))
```

```
for(i in 1:length(outcome))
{
sta1[i,1]=length(which(outcome[[i]]$abund[1,]!=0))
sta1[i,2]=length(which(outcome[[i]]$abund[2,]!=0))
if(sta1[i,1]<sta1[i,2])
{
sta1[i,3]=2
}
if(sta1[i,1]>sta1[i,2])
{
sta1[i,3]=1
}
if(sta1[i,1]==sta1[i,2])
{
sta1[i,3]=0
}
}#end for
return(sta1)
}

#function for checking bare islands in the simulation
sta.baresite<-function(outcome,island)
{
sitenum=island
sta<-matrix(0,ncol=3,nrow=length(outcome))
s1=0
s2=0
s12=0
for(i in 1:length(outcome))
{
c1=which(outcome[[i]]$abund[1,]==0)
c2=which(outcome[[i]]$abund[2,]==0)
if(length(c1)!=0)
{
if(c1[1]!=1 & c1[1]+length(c1)-1<sitenum)
{
sta[i,1]=1
}
}
```

```
if(length(c2)!=0)
{
if(c2[1]!=1 & c2[1]+length(c2)-1<sitenum)
{
sta[i,2]=1
}
}

if(sta[i,1]==1 & sta[i,2]==1)
{
s12=s12+1
}
if(sta[i,1]==0 & sta[i,2]==1)
{
s2=s2+1
}
if(sta[i,1]==1 & sta[i,2]==0)
{
s1=s1+1
}
}#end for

return(list(equal=s12,s1bare=s1,s2bare=s2))
}

#function for species abundance rank analysis
sta.rank<-function(outcome)
{
sta<-matrix(0,ncol=3,nrow=length(outcome))
for(i in 1:length(outcome))
{
sta[i,1]=sum(outcome[[i]]$abund[1,])
sta[i,2]=sum(outcome[[i]]$abund[2,])
if(sta[i,1]>sta[i,2])
{
sta[i,3]=1
}
if(sta[i,1]<sta[i,2])
{
```

```
sta[i,3]=2
}
if(sta[i,1]==sta[i,2])
{
sta[i,3]=0
}
}#end for
return(sta)
}#end function

#function for coexistence analysis for two species
sta.coexistence<-function(outcome,islandnum=island)
{
sta<-matrix(0,ncol=3+length(outcome[[1]]$pars),nrow=length(outcome))
for(i in 1:length(outcome))
{
s12=0
s1=0
s2=0
for(j in 1:island)
{
if(outcome[[i]]$abund[1,j]!=0 & outcome[[i]]$abund[2,j]!=0)
{
s12=s12+1
}
if(outcome[[i]]$abund[1,j]!=0 & outcome[[i]]$abund[2,j]==0)
{
s1=s1+1
}
if(outcome[[i]]$abund[1,j]==0 & outcome[[i]]$abund[2,j]!=0)
{
s2=s2+1
}
}#end for j
sta[i,1]=s1
sta[i,2]=s2
sta[i,3]=s12-1 #not consider the first site
sta[i,4:dim(sta)[2]]=outcome[[i]]$pars
```

```
colnames(sta)<-
c("s1win","s2win","coexist","r1","r2","disp","comp1","comp2")
}#end i
return(sta)
}#end function

# function for coexistence analysis for multiple species
sta.mcoexistence<-function(outcome,island=island,spnum=spnum)
{
conum<-spnum*2 #s1,s2,s3,two,three
sta<-
matrix(0,ncol=conum+length(outcome[[1]]$pars),nrow=length(outcome))
for(i in 1:length(outcome))
{
for(j in 2:island)#not consider the first patch
{
num<-length(which(outcome[[i]]$abund[,j]!=0))
if(num==0)
{
sta[i,1]=sta[i,1]+1
}
if(num==1)
{
spindex<-which(outcome[[i]]$abund[,j]!=0)
sta[i,spindex+1]=sta[i,spindex+1]+1
}
if(num>1)
{
sta[i,spnum+num]=sta[i,spnum+num]+1
}
}
sta[i,(conum+1):dim(sta)[2]]=as.numeric(outcome[[i]]$pars)    #important
to make a transformation
}#end i
colnames(sta)<-
c(paste("s",c(0:spnum),sep=""),paste("co",c(2:spnum),sep=""),names(outcome
[[i]]$pars))
return(sta)
}#end function
```

```
#function for neutral versus niche cases
sta.fitness<-function(coexistence)
{
neutral<-coexistence[which(coexistence[,4]==coexistence[,5]),]
niche<-coexistence[which(coexistence[,4]!=coexistence[,5]),]
neutral.num<-dim(neutral)[1]
niche.num<-dim(niche)[1]
conum<-matrix(0,ncol=2,nrow=island)
colnames(conum)<-c("neutral","niche")
for(i in 1:(island-1))
{
conum[i,1]<-length(which(neutral[,3]==island-i))
conum[i,2]<-length(which(niche[,3]==island-i))
}
conum[island,1]=neutral.num
conum[island,2]=niche.num
return(conum)
}

#function for different parameter rate comparison
sta.comparison<-function(coexistence,parameters=parspace)
{
comparisonlist<-list()
length(comparisonlist)<-dim(coexistence)[2]-3
conum<-matrix(0,ncol=length(parameters)+1,nrow=island)

for(pars in 1:length(comparisonlist))
{
for(i in 1:(island-1))
{
conum[i,1]=island-i
for(j in 1:length(parameters))
{
conum[i,j+1]<-length(which(coexistence[,3]==island-i&
coexistence[,pars+3]==parameters[j]))
}#end j
}#end i
for(j in 1:length(parameters))
{
```

```
conum[island,j+1]=length(which(coexistence[,pars+3]==parameters[j]))
}
comparisonlist[[pars]]<-conum
}
names(comparisonlist)<-colnames(coexistence[,4:dim(coexistence)[2]])
return(comparisonlist)
}

#function for different parameter rate comparison for multiple species
sta.mcomparison<-
function(coexistence,coenum,island,spnum,parameters=parspace)
{
comparisonlist<-list()
length(comparisonlist)<-dim(coexistence)[2]-2*spnum
conum<-matrix(0,ncol=length(parameters)+1,nrow=island)
colnames(conum)<-c("coe.num",paste("=",parameters,sep=""))

for(pars in 1:length(comparisonlist))
{
for(i in 1:(island-1))
{
conum[i,1]=island-i #number of pacthes for which there are coenum
species coexist
for(j in 1:length(parameters))
{
conum[i,j+1]<-length(which(coexistence[,spnum+coenum]==island-i
&coexistence[,pars+2*spnum]==parameters[j]))
}#end j
}#end i
for(j in 1:length(parameters))
{
conum[island,j+1]=length(which(coexistence[,spnum+coenum]==paramet
ers[j]))
}
comparisonlist[[pars]]<-conum
}
names(comparisonlist)<-
colnames(coexistence[,(2*spnum+1):dim(coexistence)[2]])
return(comparisonlist)
```

```
}

#function for pairwise parameter comparison
sta.paircomparison<-
function(coexistence,parnum=parnum,parameters=parspace)
{
comparisonlist<-list()
length(comparisonlist)<-parnum*(parnum-1)/2
varlist<-comparisonlist
namesvector<-vector()
length(namesvector)<-length(comparisonlist)
count=0
for(p1 in 1:(parnum-1))
{
for(p2 in (p1+1):parnum)
{
conum<-matrix(0,ncol=length(parameters),nrow=length(parameters))
varmat<-conum
count=count+1
for(i in 1:length(parameters))
{
for(j in 1:length(parameters))
{
temp<-coexistence[which(coexistence[,3+p1]==parameters[i]&
coexistence[,3+p2]==parameters[j]),]
conum[i,j]<-mean(temp[,3])
varmat[i,j]<-var(temp[,3])
}#end j
}#end i
comparisonlist[[count]]<-conum
namesvector[count]<-
paste(colnames(coexistence)[3+p1],colnames(coexistence)[3+p2],sep="-")
varlist[[count]]<-varmat
}
}
names(comparisonlist)<-namesvector
names(varlist)<-namesvector
return(list(mean=comparisonlist,var=varlist))
}#end function
```

```
#function for different parameter rate comparison
sta.comparison<-function(coexistence,parameters=parspace)
{
comparisonlist<-list()
length(comparisonlist)<-dim(coexistence)[2]-3
conum<-matrix(0,ncol=length(parameters)+1,nrow=island)
for(pars in 1:length(comparisonlist))
{
for(i in 1:(island-1))
{
conum[i,1]=island-i
for(j in 1:length(parameters))
{
conum[i,j+1]<-length(which(coexistence[,3]==island-i&
coexistence[,pars+3]==parameters[j]))
}#end j
}#end i
for(j in 1:length(parameters))
{
conum[island,j+1]=length(which(coexistence[,pars+3]==parameters[j]))
}
comparisonlist[[pars]]<-conum
}
names(comparisonlist)<-colnames(coexistence[,4:dim(coexistence)[2]])
return(comparisonlist)
}
#function for pairwise parameter comparison between multiple species
with multiple parameter #space
sta.mpaircomparison<-
function(coexistence,coenum,spnum,parameters=parspace)
{
if(coenum>spnum | coenum<=1)
{
cat("wrong number","\n")
coenum=spnum
}
parnum<-dim(coexistence)[2]-2*spnum
comparisonlist<-list()
length(comparisonlist)<-parnum*(parnum-1)/2
```

```
varlist<-comparisonlist
namesvector<-vector()
length(namesvector)<-length(comparisonlist)
count=0
for(p1 in 1:(parnum-1))
{
for(p2 in (p1+1):parnum)
{
if(!is.list(parameters))
{
conum<-matrix(0,ncol=length(parameters),nrow=length(parameters))
colnames(conum)=paste("=",parameters,sep="")
rownames(conum)=paste("=",parameters,sep="")
varmat<-conum
count=count+1
for(i in 1:length(parameters))
{
for(j in 1:length(parameters))
{
temp<-coexistence[which(coexistence[,2*spnum+p1]==parameters[i]&
coexistence[,2*spnum+p2]==parameters[j]),]
if(length(temp)==0)
{
conum[i,j]=0
varmat[i,j]=0
}else
{
conum[i,j]<-mean(temp[,spnum+coenum])
varmat[i,j]<-var(temp[,spnum+coenum])
}
}#end j
}#end i
}else #is a list of with unequal parameter spaces
{
conum<-
matrix(0,ncol=length(parameters[[p2]]),nrow=length(parameters[[p1]]))
colnames(conum)=paste("=",parameters[[p2]],sep="")
rownames(conum)=paste("=",parameters[[p1]],sep="")
varmat<-conum
```

```
count=count+1
for(i in 1:length(parameters[[p2]]))
{
for(j in 1:length(parameters[[p1]]))
{
temp<-
coexistence[which(coexistence[,2*spnum+p1]==parameters[[p2]][i]&
coexistence[,2*spnum+p2]==parameters[[p1]][j]),]
if(length(temp)==0)
{
conum[i,j]=0
varmat[i,j]=0
}else
{
conum[i,j]<-mean(temp[,spnum+coenum])
varmat[i,j]<-var(temp[,spnum+coenum])
}
}#end j
}#end i
}#list if
comparisonlist[[count]]<-conum
namesvector[count]<-
paste(colnames(coexistence)[2*spnum+p1],colnames(coexistence)[2*spnum+
p2],sep="-")
varlist[[count]]<-varmat
}
}
names(comparisonlist)<-namesvector
names(varlist)<-namesvector
return(list(mean=comparisonlist,var=varlist))
}#end function


#################################################################
###########Functions for data inputs and simulation outputs###########
#################################################################

#specific for this package
#all dataset's name are coded
```

```
#as a xxxx00yh0.xxxx.dat format

#function for reading simulation output data from a specific path
read.patchdata<-function(path=NULL,spnum=2,islandnum=island)
{
if(length(path)!=0)
{
raw<-scan(path,what=character(),sep="\t")
fileline<-length(count.fields(path))
outlist<-list()
length(outlist)<-fileline/(spnum+1)

sp<-matrix(0,nrow=spnum,ncol=islandnum)

count=0
for(i in 1:length(raw))
{
if(raw[i]=="V1")
{
count=count+1
for(j in 1:spnum)
{
sp[j,]=as.numeric(raw[(i+10+j+(j-1)*islandnum):(i+10+j-
1+j*islandnum)])
}
outlist[[count]]<-list(abund=sp)
}
}
}
return(outlist)
}#end function

#function for reading data with index
read.data<-
function(path=NULL,index=NULL,spnum=2,islandnum=island)
{
if(is.character(index))
{
indmat<-read.table(index,header=T)
```

```
index=indmat
}

if(length(path)!=0)
{
raw<-scan(path,what=character(),sep="\t")
fileline<-length(count.fields(path))
outlist<-list()
length(outlist)<-fileline/(spnum+1)

sp<-matrix(0,nrow=spnum,ncol=islandnum)

count=0
for(i in 1:length(raw))
{
if(raw[i]=="V1")
{
count=count+1
for(j in 1:spnum)
{
sp[j,]=as.numeric(raw[(i+10+j+(j-1)*islandnum):(i+10+j-
1+j*islandnum)])
}

#read pars
par<-index[count,] #save parameters
names(par)<-
c(paste("r",1:spnum,sep=""),paste("dis",1:spnum,sep=""),paste("c",1:spnum,se
p=""))
#save both abundance and pars data
outlist[[count]]<-list(abund=sp,pars=par)
}
}
}
return(outlist)
}#end function

#function for batch reading different file data based on the order 1, 2, 3, 4,
5, 6...
```

```
#it can directly read data from the raw file names without truncation
batch.read<-
function(path=pathvector,index=NULL,spnum=2,islandnum=island)
{
num<-length(path)
outlist<-list()
length(outlist)<-num
if(length(index)==0)
{
for(i in 1:num)
{
outlist[[i]]<-
read.patchdata(path=path[i],spnum=spnum,islandnum=islandnum)
}
}
if(length(index)!=0)
{
for(i in 1:num)
{
outlist[[i]]<-read.data(path=path[i],index=index,
spnum=spnum,islandnum=islandnum)
}
}
return(outlist)
}

#convert saved data sets' names in to a list
#based on the order of numbers
convert.filenames<-function(folder)
{
files<-list.files(path=folder,full.names=T)
newf<-vector()
fnum<-length(files)
length(newf)<-fnum
dataorder<-rep(0,1,fnum)
for(i in 1:fnum)
{
pos.end<-unlist(gregexpr("00yh",files[i]))[1]-1
pos.start<-unlist(gregexpr(paste(folder,"/out",sep=""),files[i]))[1]+15
```

```
dataorder[i]<-as.numeric(substr(files[i],pos.start,pos.end))
}
for(i in 1:fnum)
{
newf[dataorder[i]]<-files[i]
}
newf<-newf[!is.na(newf)]
return(newf)
}

#transform number vector into transition type of variables
trans<-function(modelnum)
{
h<-vector()
length(h)<-length(modelnum)
for(i in 1:length(modelnum))
{
h[i]<-sampletype(modelnum[i],modelnum[i])
}
return(h)
}
#

#open and create a new folder if not existed
filename.check<-function(path=NULL,return=TRUE)
{
randnum<-runif(1)
if(length(path)!=0)
{
pos<-unlist(gregexpr("/",path))
dot<-unlist(gregexpr(".",path,fixed=T))
dd<-unlist(gregexpr(":",path,fixed=T))
special<-unlist(gregexpr("00yh",path))
if(length(pos)>=2 & dot[1]!=-1 & dd[1]!=-1 & special[1]!=-1)
{
folder<-substr(path,1,pos[length(pos)]-1)
dir.create(folder,showWarnings=F)
fname<-path
}
```

```
if(length(pos)>=2 & dot[1]!=-1 & dd[1]!=-1 & special[1]==-1)
{
folder<-substr(path,1,pos[length(pos)]-1)
dir.create(folder,showWarnings=F)
fname<-paste(substr(path,1,dot[length(dot)]-
1),"00yh",substr(path,dot[length(dot)],nchar(path)),sep="")
}
if(dot[1]==-1 & dd[1]!=-1 & special[1]==-1)
{
dir.create(path,showWarnings=F)
fname<-paste(path,"/",randnum,"00yh",".dat",sep="")
}
if(dot[1]==-1 & dd[1]==-1 & special[1]==-1)
{
path<-paste("c://",path,sep="")
dir.create(path,showWarnings=F)
fname<-paste(path,"/",randnum,"00yh",".dat",sep="")
}
}
if(length(path)==0)
{
if(length(folder)==0)
{folder="c://outcome"}
fname<-paste(folder,"/",randnum,"00yh",".dat",sep="")
dir.create(folder,showWarnings=F)
}
if(return==T)
{
return(fname)
}
}#end check


#######################################################################
################Functions for figure plotting#########################
#######################################################################
```

```
#make a heatmap based on matrix values
make.heatmap<-
function(mat,type="gray",xname="x",yname="y",xlab=NULL,ylab=NULL,titl
e=NULL)
{
xrange<-dim(mat)[2]+1
yrange<-dim(mat)[1]+1
if(xrange>=yrange)
{
maxrange<-xrange
xscale=1
yscale<-yrange/xrange
}
if(yrange>xrange)
{
maxrange<-yrange
xscale<-xrange/yrange
yscale=1
}
maxnum=ceiling(max(as.vector(mat)))
tmat<-t(mat)
plot((1:maxrange)*xscale,(1:maxrange)*yscale,type="n",xlab=xname,ylab
=yname,axes=F)
if(length(xlab)!=0)
{
axis(1,at=c(1:(xrange-1))+.5,labels=xlab)
}else
{
axis(1,at=c(1:(xrange-1))+.5,labels=c(1:(xrange-1)))
}
if(length(ylab)!=0)
{
axis(2,at=c(1:(yrange-1))+.5,labels=ylab)
}else
{
axis(2,at=c(1:(yrange-1))+.5,labels=c(1:(yrange-1)))
}
if(length(title)!=0)
{
```

```
mtext(title)
}
for(i in 1:(xrange-1))
{
for(j in 1:(yrange-1))
{
value=tmat[i,j]
x1=i
x2=i+1
y1=j
y2=j+1
if(type=="gray")
{
rect(x1,y1,x2,y2,col=gray((maxnum-
value)/maxnum),border=gray((maxnum-value)/maxnum))
}
}#j
}#i
}#end function


####################################################################
#####################Other auxiliary functions######################
####################################################################

#function for constructing a full list of all combinations of parameters
#parvecotr is a parameter vector, vector
#parnum is the number of parameters, numeric
comblist<-function(parvector,parnum)
{
combnum<-length(parvector)^parnum
mat<-matrix(0,ncol=parnum,nrow=combnum)
for(i in 1:parnum)
{
leg<-length(parvector)^(parnum-i)
period=length(parvector)^i
repeated=length(parvector)^(i-1)
fullcircle=leg*length(parvector)
temp<-vector()
```

```
length(temp)<-fullcircle
for(k in 1:length(parvector))
{
temp[((k-1)*leg+1):(k*leg)]=parvector[k]
}

for(ii in 1:repeated)
{
mat[((ii-1)*fullcircle+1):(ii*fullcircle),i]=temp
}
}
return(mat)
}#end function

# function for a full list of all combinations of parameters
#parvecotr is a parameter list, can vary on size
#parnum is the number of parameters, numeric
comblist2<-function(parlist)
{
combnum=1
parnum<-length(parlist)
lvector<-vector()
length(lvector)<-parnum
for(i in 1:parnum)
{
combnum<-length(parlist[[i]])*combnum
lvector[i]=length(parlist[[i]])
}
mat<-matrix(0,ncol=parnum,nrow=combnum)
for(i in 1:parnum)
{
leg<-prod(lvector[i:parnum])/lvector[i]
repeated=combnum/prod(lvector[i:parnum])
fullcircle=leg*lvector[i]
temp<-vector()
length(temp)<-fullcircle
for(k in 1:length(parlist[[i]]))
{
temp[((k-1)*leg+1):(k*leg)]=parlist[[i]][k]
```

```
}

for(ii in 1:repeated)
{
mat[((ii-1)*fullcircle+1):(ii*fullcircle),i]=temp
}
}
return(mat)
}#end function


#####################################################################
#################Some annotations about the package#################
#####################################################################

#1, dispersal rate must be set as small as possible
# because dispersal is always rare,otherwise the transition
# can be easily reached
#2, coexistence under unstable saddle points can be possible
# as dispersal serve as the ceaseless source to supply species
# this process can overcome the competition exclusion
#3, interestingly, coexistence of both species can be the
# outcome of stable exclusive points transitions
# for example, look at the parameter set
# c(0.9 0.9 0.5       0.25     0.5). Transition from stage (S1=4000,S2=0)
# to (S1=0,S2=1000) from good to bad habitats, will lead to
# an unstable coexistence scenario at site 7 as long as dispersal always happen
#4, why species extinction goes fast? Dipersal fragmentation and discontinuity
# is the reason to drive unstable coexistence scenarios to disappear
#5, for fitness equivalently cases (neutral process), we found that
# the coexistence is largely promoted


#####################################################################
########################Practical simulation section##################
#####################################################################
```

```
#basic important parameters for configuration of the simulation
island=10 #patch number
spnum=2 #species number
parnum=6 #parameter number=species number*3
good=1000 #habitat quality value-high
bad=500 #habitat quality value-low
scale=2 #variable shifting from different patches
dispersalscale=51 #specific scale for dispersal rates
folder="c://outcome" #global folder to save all the data and figures


##################################################################
########Different simulation scenarios with different conditions#########
##################################################################

out<-list()
parspace<-c(.1,.25,.5,.75,.9)
h<-vector()
length(h)<-modelnum
for(i in 1:modelnum)
{
h[i]<-sampletype(i,i)
}

#decreasing habitats
t<-c(h[1],h[1],h[3],h[3],h[3],h[3],h[3])
out[[1]]<-
sim.coarse(scale=2,allee=1,T=500,prange=parspace,type=t,path="c://outcome/
out1")
t<-c(h[1],h[1],h[3],h[3],h[1],h[3],h[3])
out[[2]]<-
sim.coarse(scale=2,allee=1,T=500,prange=parspace,type=t,path="c://outcome/
out2")

#mosaic habitats
t<-c(h[4],h[4],h[3],h[3],h[3],h[3],h[3])
out[[3]]<-
sim.coarse(scale=2,allee=1,T=500,prange=parspace,type=t,path="c://outcome/
out3")
```

```
    t<-c(h[4],h[4],h[3],h[3],h[4],h[3],h[3])
    out[[4]]<-
sim.coarse(scale=2,allee=1,T=500,prange=parspace,type=t,path="c://outcome/
out4")

    #distinct habitats1
    t<-c(h[1],h[4],h[3],h[3],h[3],h[3],h[3])
    out[[5]]<-
sim.coarse(scale=2,allee=1,T=500,prange=parspace,type=t,path="c://outcome/
out5")
    t<-c(h[1],h[4],h[3],h[3],h[4],h[3],h[3])
    out[[6]]<-
sim.coarse(scale=2,allee=1,T=500,prange=parspace,type=t,path="c://outcome/
out6")
    #

    #increasing habitats
    t<-c(h[2],h[2],h[3],h[3],h[3],h[3],h[3])
    out[[11]]<-
sim.coarse(scale=2,allee=1,T=500,prange=parspace,type=t,path="c://outcome/
out11")
    t<-c(h[2],h[2],h[3],h[3],h[2],h[3],h[3])
    out[[12]]<-
sim.coarse(scale=2,allee=1,T=500,prange=parspace,type=t,path="c://outcome/
out12")

    #distinct habitats3
    t<-c(h[2],h[4],h[3],h[3],h[3],h[3],h[3])
    out[[13]]<-
sim.coarse(scale=2,allee=1,T=500,prange=parspace,type=t,path="c://outcome/
out13")
    t<-c(h[2],h[4],h[3],h[3],h[4],h[3],h[3])
    out[[14]]<-
sim.coarse(scale=2,allee=1,T=500,prange=parspace,type=t,path="c://outcome/
out14")
    t<-c(h[2],h[4],h[3],h[3],h[2],h[3],h[3])
    out[[15]]<-
sim.coarse(scale=2,allee=1,T=500,prange=parspace,type=t,path="c://outcome/
out15")
```

```
#others…

#changing a pair variable simultaneously across habitats
t<-c(h[1],h[1],h[3],h[3],h[3],h[1],h[1])
out[[16]]<-
sim.coarse(scale=2,allee=1,T=500,prange=parspace,type=t,path="c://outcome/
out16")
t<-c(h[1],h[1],h[1],h[1],h[3],h[3],h[3])
out[[17]]<-
sim.coarse(scale=2,allee=1,T=500,prange=parspace,type=t,path="c://outcome/
out17")

#changing one of a pair variable across habitats
t<-c(h[1],h[1],h[3],h[3],h[3],h[3],h[1])
out[[18]]<-
sim.coarse(scale=2,allee=1,T=500,prange=parspace,type=t,path="c://outcome/
out18")
t<-c(h[1],h[1],h[1],h[3],h[3],h[3],h[3])
out[[19]]<-
sim.coarse(scale=2,allee=1,T=500,prange=parspace,type=t,path="c://outcome/
out19")
#

#species can have differentiated dispersal ability
#similar to model 1 and 2
t<-c(h[1],h[1],h[3],h[3],h[3],h[1],h[3],h[3])
out[[20]]<-
fast.flexsim(scale=2,allee=1,T=400,parcomb=allcomb,spnum=2,sourcetype="
constant",type=t,path="c://outcome/out20")
#similar to model 1 and 2,high negative covariance between habitat and
dispersal
t<-c(h[1],h[2],h[3],h[3],h[2],h[1],h[3],h[3])
out[[21]]<-
fast.flexsim(scale=2,allee=1,T=400,parcomb=allcomb,spnum=2,sourcetype="
constant",type=t,path="c://outcome/out21")
#similar to model 1,2, but source type is modified
t<-c(h[1],h[1],h[3],h[3],h[3],h[3],h[3],h[3])
```

```
out[[22]]<-
fast.flexsim(scale=2,allee=1,T=400,parcomb=allcomb,spnum=2,sourcetype="f
lexible",type=t,path="c://outcome/out22")

#similar to model 1,2, but totally homogeneous
t<-c(h[3],h[3],h[3],h[3],h[3],h[3],h[3])
out[[23]]<-
sim.coarse(scale=2,allee=1,T=500,prange=parspace,type=t,path="c://outcome/
out23")

#similar to model 8,9,10, but dispersal is modified
t<-c(h[1],h[5],h[3],h[3],h[1],h[4],h[3],h[3])
out[[24]]<-
fast.flexsim(scale=2,allee=1,T=400,parcomb=allcomb,spnum=2,sourcetype="
constant",type=t,path="c://outcome/out24")

#similar to model 11-15, but dispersal rates between species are modified
t<-c(h[4],h[2],h[3],h[3],h[5],h[1],h[3],h[3])
out[[25]]<-
fast.flexsim(scale=2,allee=1,T=400,parcomb=allcomb,spnum=2,sourcetype="
constant",type=t,path="c://outcome/out25")

#similar to model 16-19, but all parameters varied rates negatively related
to habitat quality
t<-c(h[5],h[2],h[4],h[1],h[4],h[1],h[4],h[1])
out[[26]]<-
fast.flexsim(scale=2,allee=1,T=400,parcomb=allcomb,spnum=2,sourcetype="
constant",type=t,path="c://outcome/out26")

#similar to model 16-19, but all parameters varied rates negatively related
to habitat quality
t<-c(h[4],h[2],h[5],h[1],h[5],h[1],h[5],h[1])
out[[27]]<-
fast.flexsim(scale=2,allee=1,T=400,parcomb=allcomb,spnum=2,sourcetype="
constant",type=t,path="c://outcome/out27")
t<-c(h[1],h[2],h[2],h[1],h[2],h[1],h[2],h[1])
out[[28]]<-
fast.flexsim(scale=2,allee=1,T=400,parcomb=allcomb,spnum=2,sourcetype="
constant",type=t,path="c://outcome/out28")
```

```
t<-c(h[1],h[3],h[2],h[3],h[2],h[3],h[2],h[3])
out[[29]]<-
fast.flexsim(scale=2,allee=1,T=400,parcomb=allcomb,spnum=2,sourcetype="
constant",type=t,path="c://outcome/out29")
t<-c(h[1],h[3],h[2],h[3],h[2],h[3],h[2],h[3])
out[[30]]<-
fast.flexsim(scale=2,allee=1,T=400,parcomb=allcomb,spnum=2,sourcetype="f
lexible",type=t,path="c://outcome/out30")
#
```

AUTHOR'S CONTACT INFORMATION

Dr. Youhua Chen, PhD
Department of Renewable Resources
University of Alberta, Edmonton, T6G 2H1, Canada
Email: haydi@126.com

INDEX

A

adjusted R^2, 51, 101, 103
administrative units, 75, 76, 84
akaike information criteria, 80
alien species, 98, 109
alpha diversity, 29, 30, 31
amensalism, 135
ancestral range reconstruction, 56
ancestral ranges, 55, 56, 57, 59, 60, 61, 62, 63, 64, 65
annual evaporation, 104
asymmetric connectivity, 120, 136
asymmetric dispersal, 112, 118, 120, 134, 137
asymmetric processes, 119
asymptotic normal distribution, 3, 5
AUC, 99, 102

B

beta diversity, 8, 12, 13, 14, 17, 18, 19, 20, 21, 22, 23, 24, 25, 26, 27, 29, 30, 31, 32, 33, 34, 37, 38, 39, 46, 50, 52, 53, 108
binary range transition model, 62
binary state transition matrix, 57
biodiversity hotspots, 134
biodiversity maintenance, vii, 111, 117, 119, 132
biodiversity resources, vii, 26
biogeographic inferences, 57
biological processes, 68
biotic interaction, 48, 112, 128
boundary edges, 135

C

canonical correspondence analysis (CCA), 32, 37, 45, 46, 97, 101, 105
canopy arthropods, 87, 90, 91, 92
carrying capability, 58, 62
CCA-based variation partitioning, 37
clade age, 68, 70, 73
clade age-species richness relationship, 68
clade history, 67
climate change, 94, 98, 101, 102, 103, 104
climatic condition, 76, 83, 84, 104
climatic factor, 103
climatic fluctuation, 78
climatic layers, 99, 100
climatic variability, 76, 80, 82, 83
climatic warning, 104
clumped dispersal, 120
coexistence, 52, 111, 112, 113, 118, 119, 120, 121, 123, 125, 126, 128, 129, 130, 131, 132, 133, 134, 135, 137, 145, 151, 153, 154, 155, 156, 170, 171, 172, 173, 174, 175, 176, 177, 186, 187
coexistence outcomes, 120, 121, 122, 129
colonization, 48, 64, 65, 104, 115, 121, 137

community ecology, 3, 8, 30, 41, 46, 137
community stability, 13
community structures, 49
community turnover, 11, 14, 15, 32, 33, 36
community-level diversity, 14, 29, 31
competition, 48, 83, 85, 111, 114, 115, 117, 119, 120, 121, 122, 123, 124, 126, 127, 128, 130, 131, 133, 134, 135, 136, 137, 148, 150, 151, 152, 153, 160, 162, 163, 165, 167, 187
competition coefficients, 114, 123, 126, 128
competition exclusions, 134
complete spatial randomness (CSR), 77, 79, 81
compositional variation, 13, 31
computational loads, 4
connectivity, 44, 58, 62, 113, 114, 115, 120, 125, 127, 128, 136
connectivity matrix, 125
contemporary distribution, 76
contour plot, 64, 112
correlation analysis, 37
corridor hotspots, 134, 135
county locations, 78, 80, 82, 83
critically endangered, 104
cropland density, 78, 80, 83
cropland distribution, 77

D

data uncertainty, 87, 90, 92, 93
deleting process, 12, 17, 18, 20, 25
density, 32, 37, 50, 79, 83, 85, 116, 130, 135, 136, 155, 156
density dependence, 130, 135
discordant, 83
dispersal ability, 38, 43, 111, 112, 114, 123, 125, 128, 190
dispersal rate(s), 58, 59, 62, 63, 64, 115, 125, 127, 128, 131, 133, 134, 135, 163, 165, 186, 187, 190
dissimilarity, 26, 37
distributional points, 79
distributional ranges, 32, 58, 64, 97
diversification rates, 67, 69, 70, 72, 73

E

ecological communities, 12, 13, 26, 37, 51, 52, 136
ecological laws, 88
ecological niche modeling, 98
ecological opportunity, 104
ecological scaling, 88
ecosystem functioning, 12, 27, 30
edge effects, 49
edges of habitats, 135
empirical test, 43, 124
endemic reptiles, 99
environment, 6, 29, 31, 32, 35, 36, 43, 51, 73, 80, 85, 93, 97, 111, 115, 137
environmental filtering, 30, 41, 49, 50, 119
environmental variables, 43, 44, 49
epidemic diseases, 98
epigean arthropods, 87, 90, 91, 92
estimation uncertainties, 101
euclidean distances, 48
evolutionary history, 68, 72
evolutionary models, 56, 64
evolutionary time, 67, 69, 70, 73
extinction risk, 30, 88, 107, 120

F

fitness equivalence, 120, 129, 131
fluctuating environments, 112, 120
flying ability, 121
forest birds, 46
forward variable selection, 50
future climatic scenarios, 99, 102, 103

G

general additive models, 41, 45, 51
geographic coordinates, 48, 78
geographic information system, 77, 78
geometric constraints, 121
G-function graph, 81
goodness of fit, 90, 92
ground bryophytes, 87, 91, 92

H

habitat conditions, 125
habitat heterogeneity, 98
heatmap, 112, 116, 117, 157, 158, 159, 183
heteroscedasticity, 47, 48
historical biogeographic patterns, 64
historical biogeography, 56, 65
historical influence, 83
historical residence, 76
homogeneity, 34
homogeneous poisson process, 79
homogenizing process, 11, 17, 18, 19, 23
human activities, 77, 83
human disturbance, 30, 31, 34
human footprint index, 77, 78, 80
human population size, 76
human society, 76

I

immigration matrix, 126
immigration rates, 130
industrialization, 76
inhomogeneous poisson process, 79
initialization, 126, 145
internal node, 59, 60, 63, 68, 72
intraspecific competition, 122, 126, 130, 131
intrinsic per capita growth rates, 126

L

landscape connectivity, 120
landscape heterogeneity, 78, 88
Lasso, 50, 53
local dispersal, 120
Log-likelihood, 61
log-transformation, 90, 93, 96
Lotka-Volterra competition model, 122
lower triangular, 5

M

macroecology, 67
macroevolution, 67
mainland China, vi, 75, 76, 77, 78, 80, 81, 82, 83, 84
mammalian phylogeny, 68, 69
mantel statistic, 3, 4, 5, 6, 7, 8, 142
mantel test-based variation partitioning, 37
matrix comparison, 3
MaxEnt, 99, 102
maximal temperature, 80, 104
maximum likelihood, 55, 56, 59, 60, 64, 73
mean temperature, 78, 100, 104
measurement errors, 88, 93
measurement uncertainty, 87, 88, 89, 92
mega-diversity, 72
minimal temperature, 100, 104
model configuration, 113, 115
Monte Carlo method, 56
multifaceted factors, 76
multiple regression analysis, 44, 101, 103
multiples, 26
multiple-site comparison, 12, 31, 33, 34
multivariate regression, 32, 42
multivariate statistics, 41, 42
mutualism, 135, 136

N

nearest neigbhor, 78
nestedness, 11, 12, 13, 14, 15, 16, 17, 18, 20, 26, 27, 30, 31, 33, 34, 36, 37, 38, 39
neutral process, 39, 43, 45, 53, 119, 129, 131, 132, 135, 145, 187
neutrality, 43, 45, 46, 49, 119, 120, 129, 131, 132
niche, 30, 39, 43, 45, 46, 49, 51, 53, 97, 98, 108, 109, 112, 119, 120, 122, 129, 131, 132, 135, 145, 154, 172
niche process, 45, 49, 51, 119, 122, 129, 131, 132, 145
nonlinear weighted least-squares, 89, 92, 93

null hypothesis, 5, 7, 8
numerical simulations, 4, 11, 112

O

occupancy-abundance relationship, 97, 98, 99, 105
occurrence probability, 97, 99, 101, 105
optimization, 89, 93, 112
ordinary least-square regression, 69, 71
ordination, 31, 37, 41, 45, 47, 101
oribatid mites, 46

P

pair-site comparison, 33, 34
partial ordination, 99
partial regression, 49
partitioning of beta-diversity, 20
Pelagic fish, 46, 52
permutation, 3, 4, 6, 7
phylogenetic comparative methods, 67
phylogenetic distance, 68
phylogenetic history, 64, 73
phylogenetic tree, 61, 65
phylogenetical general regression analysis, 69
phylogeographic studies, 57
point randomness, 78
polynomial regression, 48
population declining, 103, 104
population size, 76, 77, 83, 122, 124, 125, 133
power-law, 87, 90, 92, 93
predation, 135
prediction accuracy, 50
presence/absence, 55, 105
probability of occurrence, 100, 105
pseudo-communities, 12, 18, 19, 21, 22
psychotria, 55, 60, 63, 64, 65

Q

quadratic nonlinear models, 48
quantile, 5

R

R package, 9, 14, 52, 69, 84, 111
random sampling, 100
randomness, 3
range contraction, 102
range evolution, 55, 56, 57, 58, 59, 60, 61, 64, 65
range evolution matrix, 59
range rarity, 32
range sizes, 103
rarity, 17, 29, 30, 32, 34, 37, 38, 39
rate matrix, 58, 146
rate-shifting models, 69
realistic scenario, 133
redundancy analysis, 32, 45
regression, 9, 31, 38, 41, 42, 43, 44, 45, 46, 47, 48, 49, 51, 67, 69, 70, 71, 96, 101, 103
removing/adding process, 12, 18, 20, 21, 22, 24
replacement, 11, 13, 16, 17, 18, 20, 26, 29, 31, 33, 34, 36, 37, 38
resource partitioning, 121, 123, 127, 128, 134, 135
richness difference, 11, 14, 16, 18, 20, 26, 29, 31, 33, 34, 36, 37, 38
ridge regression, 50

S

sampling biases, 5
sampling effect, 49, 98
sampling plots, 17, 32, 35, 50
significant level, 5
similarity index, 14, 32
simulation time, 112, 114, 115, 128
simulation-free, 4
simultaneous coupling, 121
site independence, 57
small island effect, 88, 96
socioeconomic processes, 76

socioeconomic variables, 78
source-sink system, 120, 131
space, 29, 30, 31, 32, 35, 36, 43, 44, 97, 99, 101, 103, 106, 109, 114, 115, 116, 117, 120, 124, 161, 164, 167, 175
space variation, 43, 44
spatial autocorrelation, 6, 30, 46, 98, 107
spatial coordinates, 44, 124
spatial descriptors, 43, 44
spatial limitation, 49
spatial point analysis, 77
spatial scale, 46, 50, 52, 65, 75, 83, 84, 122, 124
spatial statistic, 75, 78
spatially implicit, 121
species abundance, 16, 48, 97, 99, 100, 101, 102, 104, 107, 108, 126, 132, 136, 169
species coexistence, 111, 112, 116, 118, 119, 120, 121, 122, 126, 132, 133, 135, 136, 137, 156
species extinction, 12, 30, 95, 134, 187
species growth rates, 111, 114
species loss, 17, 31, 36, 37, 131
species occupancy, 98, 100, 101, 103, 105, 108
species occupancy change, 101
species richness, 26, 38, 67, 68, 69, 70, 71, 72, 73, 74, 87, 88, 89, 90, 91, 92, 93, 94, 95, 109
species traits, 112
species-area relationship, 85, 88, 94, 95, 96
species-environment relationships, 97
species-site matrices, 8
standard deviation, 5, 89, 91, 92, 101
statistical inference, 88
stream fish, 46
subsidizing mechanism, 119, 132, 133

T

temporal shifting, 69
terrestrial ecosystems, 45, 46
theoretical simulation, 135
time-reversible, 58, 62
total abundance, 11, 17, 34, 126
trade-off, 121, 134
transitional consequences, 135
turbulent dispersal, 120
turnover, 11, 12, 13, 14, 15, 16, 18, 20, 26, 29, 30, 31, 32, 33, 36, 37, 38

U

under-fit problem, 47
unexplained variation, 43, 44
unit normal distribution, 5
unstable coexistence, 134, 135, 187
urbanization, 75, 76, 84, 85, 86

V

variable selection, 50
variation components, 41, 44, 47
variation partitioning, 30, 39, 43, 45, 52, 53, 106

W

water availability, 76, 78, 80, 82, 84
WNODF, 11, 15, 16, 18, 20, 21, 26